Gödel's Proof

GÖDEL'S PROOF

by

Ernest Nagel

and

James R. Newman

ROUTLEDGE & KEGAN PAUL LTD

Broadway House, 68-74 Carter Lane

London EC4V 5EL

First published in England 1959
by Routledge & Kegan Paul Ltd
Broadway House, 68–74 Carter Lane
London EC4V 5EL
Reprinted 1964
Reprinted and first published
as a Routledge Paperback 1971

Printed in Great Britain
by Lowe & Brydone (Printers) Ltd, London

ISBN 0 7100 7078 0

to

Bertrand Russell

Contents

Acknowledgments

The authors gratefully acknowledge the generous assistance they received from Professor John C. Cooley of Columbia University. He read critically an early draft of the manuscript, and helped to clarify the structure of the argument and to improve the exposition of points in logic. We wish to thank *Scientific American* for permission to reproduce several of the diagrams in the text, which appeared in an article on Gödel's Proof in the June 1956 issue of the magazine. We are indebted to Professor Morris Kline of New York University for helpful suggestions regarding the manuscript.

Gödel's Proof

I

Introduction

In 1931 there appeared in a German scientific periodical a relatively short paper with the forbidding title "Uber formal unentscheidbare Sätze der Principia Mathematica und verwandter Systeme" ("On Formally Undecidable Propositions of Principia Mathematica and Related Systems"). Its author was Kurt Gödel, then a young mathematician of 25 at the University of Vienna and since 1938 a permanent member of the Institute for Advanced Study at Princeton. The paper is a milestone in the history of logic and mathematics. When Harvard University awarded Gödel an honorary degree in 1952, the citation described the work as one of the most important advances in logic in modern times.

At the time of its appearance, however, neither the title of Gödel's paper nor its content was intelligible to most mathematicians. The *Principia Mathematica* mentioned in the title is the monumental three-volume treatise by Alfred North Whitehead and Bertrand Russell on mathematical logic and the foundations of mathematics; and familiarity with that work is not a prerequisite to successful research in most branches of mathematics. Moreover, Gödel's paper deals with a set of questions that has never attracted more than a comparatively small group of students. The reasoning of the proof was so novel at the time of its publication that only those intimately conversant with the technical literature of a highly specialized field could follow the argument with ready comprehension. Nevertheless, the conclusions Gödel established are now widely recognized as being revolutionary in their broad philosophical import. It is the aim of the present essay to make the substance of Gödel's findings and the general character of his proof accessible to the non-specialist.

Gödel's famous paper attacked a central problem in the foundations of mathematics. It will be helpful to give a brief preliminary account of the context in which the problem occurs. Everyone who has been exposed to elementary geometry will doubtless recall that it is taught as a *deductive* discipline. It is not presented as an experimental science whose theorems are to be accepted because they are in agreement with observation. This notion, that a proposition may be established as the conclusion of an explicit *logical proof,* goes back to the ancient Greeks, who discovered what is known as the "axiomatic method" and used it to develop geometry in a systematic fashion. The axiomatic method consists in accepting *without* proof certain propositions as axioms or postulates (e.g., the axiom that through two points just one straight line can be drawn), and then deriving from the axioms all other propositions of the system as theorems. The axioms constitute the "foundations" of the system; the theorems are the "superstructure," and are obtained from the axioms with the exclusive help of principles of logic.

The axiomatic development of geometry made a powerful impression upon thinkers throughout the ages; for the relatively small number of axioms carry the whole weight of the inexhaustibly numerous propositions derivable from them. Moreover, if in some way the truth of the axioms can be established—and, indeed, for some two thousand years most students believed without question that they are true of space—both the truth and the mutual consistency of all the theorems are automatically guaranteed. For these reasons the axiomatic form of geometry appeared to many generations of outstanding thinkers as the model of scientific knowledge at its best. It was natural to ask, therefore, whether other branches of thought besides geometry can be placed upon a secure axiomatic foundation. However, although certain parts of physics were given an axiomatic formulation in antiquity (e.g., by Archimedes), until modern times geometry was the only branch of mathematics that had what most students considered a sound axiomatic basis.

But within the past two centuries the axiomatic method has come to be exploited with increasing power and vigor. New as well as old branches of mathematics, including the familiar arithmetic of cardinal (or "whole") numbers, were supplied with what appeared to be adequate sets of axioms. A climate of opinion was thus generated in which it was tacitly assumed that each sector of mathematical thought can be supplied with a set of axioms sufficient for developing systematically the endless totality of true propositions about the given area of inquiry.

Gödel's paper showed that this assumption is untenable. He presented mathematicians with the astounding and melancholy conclusion that the axiomatic method has certain inherent limitations, which rule out the possibility that even the ordinary arithmetic of the integers can ever be fully axiomatized. What is more, he proved that it is impossible to establish the internal logical consistency of a very large class of deductive systems—elementary arithmetic, for example—unless one adopts principles of reasoning so complex that their internal consistency is as open to doubt as that of the systems themselves. In the light of these conclusions, no final systematization of many important areas of mathematics is attainable, and no absolutely impeccable guarantee can be given that many significant branches of mathematical thought are entirely free from internal contradiction.

Gödel's findings thus undermined deeply rooted preconceptions and demolished ancient hopes that were being freshly nourished by research on the foundations of mathematics. But his paper was not altogether negative. It introduced into the study of foundation questions a new technique of analysis comparable in its nature and fertility with the algebraic method that René Descartes introduced into geometry. This technique suggested and initiated new problems for logical and mathematical investigation. It provoked a reappraisal, still under way, of widely held philosophies of mathematics, and of philosophies of knowledge in general.

The details of Gödel's proofs in his epoch-making paper are too difficult to follow without considerable mathematical training. But the basic structure of his demonstrations and the core of his conclusions can be made intelligible to readers with very limited mathematical and logical preparation. To achieve such an understanding, the reader may find useful a brief account of certain relevant developments in the history of mathematics and of modern formal logic. The next four sections of this essay are devoted to this survey.

II

The Problem of Consistency

The nineteenth century witnessed a tremendous expansion and intensification of mathematical research. Many fundamental problems that had long withstood the best efforts of earlier thinkers were solved; new departments of mathematical study were created; and in various branches of the discipline new foundations were laid, or old ones entirely recast with the help of more precise techniques of analysis. To illustrate: the Greeks had proposed three problems in elementary geometry: with compass and straight-edge to trisect any angle, to construct a cube with a volume twice the volume of a given cube, and to construct a square equal in area to that of a given circle. For more than 2,000 years unsuccessful attempts were made to solve these problems; at last, in the nineteenth century it was proved that the desired constructions are logically impossible. There was, moreover, a valuable by-product of these labors. Since the solutions depend essentially upon determining the kind of roots that satisfy certain equations, concern with the celebrated exercises set in antiquity stimulated profound investigations into the nature of number and the structure of the number continuum. Rigorous definitions were eventually supplied for negative, complex, and irrational numbers; a logical basis was constructed for the real number system; and a new branch of mathematics, the theory of infinite numbers, was founded.

But perhaps the most significant development in its long-range effects upon subsequent mathematical history was the solution of another problem that the Greeks raised without answering. One of the axioms Euclid used in systematizing geometry has to do with parallels. The axiom he adopted is logically equivalent to (though not identical with) the assumption that through a point outside a given line only one parallel to the line can be drawn. For various reasons, this axiom did not appear "self-evident" to the ancients. They sought, therefore, to deduce it from the other Euclidean axioms, which they regarded as clearly self-evident.[1] Can such a proof of the parallel axiom be

---

[1] The chief reason for this alleged lack of self-evidence seems to have been the fact that the parallel axiom makes an assertion about *infinitely remote* regions of space. Euclid defines parallel lines as straight lines in a plane that, "being produced indefinitely in both directions," do not meet. Accordingly, to say that two lines are parallel is to make the claim that the two lines will not meet even "at infinity." But the ancients were familiar with lines that, though they do not intersect each other in any finite region of the plane, do meet "at infinity." Such lines are said to be "asymptotic." Thus, a hyperbola is asymptotic to its axes. It was therefore not in-

given? Generations of mathematicians struggled with this question, without avail. But repeated failure to construct a proof does not mean that none can be found any more than repeated failure to find a cure for the common cold establishes beyond doubt that mankind will forever suffer from running noses. It was not until the nineteenth century, chiefly through the work of Gauss, Bolyai, Lobachevsky, and Riemann, that the *impossibility* of deducing the parallel axiom from the others was demonstrated. This outcome was of the greatest intellectual importance. In the first place, it called attention in a most impressive way to the fact that a *proof* can be given of the *impossibility of proving* certain propositions within a given system. As we shall see, Gödel's paper is a proof of the impossibility of demonstrating certain important propositions in arithmetic. In the second place, the resolution of the parallel axiom question forced the realization that Euclid is not the last word on the subject of geometry, since new systems of geometry can be constructed by using a number of axioms different from, and incompatible with, those adopted by Euclid. In particular, as is well known, immensely interesting and fruitful results are obtained when Euclid's parallel axiom is replaced by the assumption that more than one parallel can be drawn to a given line through a given point, or, alternatively, by the assumption that no parallels can

---

tuitively evident to the ancient geometers that from a point outside a given straight line only one straight line can be drawn that will not meet the given line even at infinity.

be drawn. The traditional belief that the axioms of geometry (or, for that matter, the axioms of any discipline) can be established by their apparent self-evidence was thus radically undermined. Moreover, it gradually became clear that the proper business of the pure mathematician is to *derive theorems from postulated assumptions,* and that it is not his concern as a mathematician to decide whether the axioms he assumes are actually true. And, finally, these successful modifications of orthodox geometry stimulated the revision and completion of the axiomatic bases for many other mathematical systems. Axiomatic foundations were eventually supplied for fields of inquiry that had hitherto been cultivated only in a more or less intuitive manner. (See Appendix, no. 1.)

The over-all conclusion that emerged from these critical studies of the foundations of mathematics is that the age-old conception of mathematics as "the science of quantity" is both inadequate and misleading. For it became evident that mathematics is simply the discipline *par excellence* that draws the conclusions logically implied by any given set of axioms or postulates. In fact, it came to be acknowledged that the validity of a mathematical inference in no sense depends upon any special meaning that may be associated with the terms or expressions contained in the postulates. Mathematics was thus recognized to be much more abstract and formal than had been traditionally supposed: more abstract, because mathematical statements can be construed in principle to be about anything whatsoever rather than about some inherently circumscribed set of objects or traits of objects; and more formal, because the validity of mathematical demonstrations is grounded in the structure of statements, rather than in the nature of a particular subject matter. The postulates of any branch of demonstrative mathematics are not inherently about space, quantity, apples, angles, or budgets; and any special meaning that may be associated with the terms (or "descriptive predicates") in the postulates plays no essential role in the process of deriving theorems. We repeat that the sole question confronting the pure mathematician (as distinct from the scientist who employs mathematics in investigating a special subject matter) is not whether the postulates he assumes or the conclusions he deduces from them are true, but whether the alleged conclusions are in fact the *necessary logical consequences* of the initial assumptions.

Take this example. Among the undefined (or "primitive") terms employed by the influential German mathematician David Hilbert in his famous axiomatization of geometry (first published in 1899) are 'point', 'line', 'lies on', and 'between'. We may grant that the customary meanings connected with these expressions play a role in the process of discovering and learning theorems. Since the meanings are familiar, we feel we understand their various interrelations, and they motivate the formulation and selection of axioms; moreover, they suggest and facilitate the formulation of the statements we hope to establish as theorems. Yet, as Hilbert plainly states, insofar as we are concerned with the primary mathematical task of exploring the purely logical relations of dependence between statements, the familiar connotations of the primitive terms are to be ignored, and the sole "meanings" that are to be associated with them are those assigned by the axioms into which they enter.[2] This is the point of Russell's famous epigram: pure mathematics is the subject in which we do not know what we are talking about, or whether what we are saying is true.

A land of rigorous abstraction, empty of all familiar landmarks, is certainly not easy to get around in. But it offers compensations in the form of a new freedom of movement and fresh vistas. The intensified formalization of mathematics emancipated men's minds from the restrictions that the customary interpretation of expressions placed on the construction of novel systems of postulates. New kinds of algebras and geometries were developed which marked significant departures from the mathematics of tradition. As the meanings of certain terms became more general, their use became broader and the inferences that could be drawn from them less confined. Formalization led to a great variety of systems of considerable mathematical interest and

---

[2] In more technical language, the primitive terms are "implicitly" defined by the axioms, and whatever is not covered by the implicit definitions is irrelevant to the demonstration of theorems.

value. Some of these systems, it must be admitted, did not lend themselves to interpretations as obviously intuitive (i.e., commonsensical) as those of Euclidean geometry or arithmetic, but this fact caused no alarm. Intuition, for one thing, is an elastic faculty: our children will probably have no difficulty in accepting as intuitively obvious the paradoxes of relativity, just as we do not boggle at ideas that were regarded as wholly unintuitive a couple of generations ago. Moreover, as we all know, intuition is not a safe guide: it cannot properly be used as a criterion of either truth or fruitfulness in scientific explorations.

However, the increased abstractness of mathematics raised a more serious problem. It turned on the question whether a given set of postulates serving as foundation of a system is internally consistent, so that no mutually contradictory theorems can be deduced from the postulates. The problem does not seem pressing when a set of axioms is taken to be about a definite and familiar domain of objects; for then it is not only significant to ask, but it may be possible to ascertain, whether the axioms are indeed true of these objects. Since the Euclidean axioms were generally supposed to be true statements about space (or objects in space), no mathematician prior to the nineteenth century ever considered the question whether a pair of contradictory theorems might some day be deduced from the axioms. The basis for this confidence in the consistency of Euclidean geometry is the sound principle that logically incompatible statements cannot be simultaneously true; accordingly, if a set of statements is true (and this was assumed of the Euclidean axioms), these statements are mutually consistent.

The non-Euclidean geometries were clearly in a different category. Their axioms were initially regarded as being plainly false of space, and, for that matter, doubtfully true of anything; thus the problem of establishing the internal consistency of non-Euclidean systems was recognized to be both formidable and critical. In Riemannian geometry, for example, Euclid's parallel postulate is replaced by the assumption that through a given point outside a line *no* parallel to it can be drawn. Now suppose the question: Is the Riemannian set of postulates consistent? The postulates are apparently not true of the space of ordinary experience. How, then, is their consistency to be shown? How can one prove they will not lead to contradictory theorems? Obviously the question is not settled by the fact that the theorems already deduced do not contradict each other—for the possibility remains that the very next theorem to be deduced may upset the apple cart. But, until the question is settled, one cannot be certain that Riemannian geometry is a true alternative to the Euclidean system, i.e., equally valid mathematically. The very possibility of non-Euclidean geometries was thus contingent on the resolution of this problem.

A general method for solving it was devised. The underlying idea is to find a "model" (or "interpretation") for the abstract postulates of a system, so that each postulate is converted into a true statement about the model. In the case of Euclidean geometry, as we have noted, the model was ordinary space. The method was used to find other models, the elements of which could serve as crutches for determining the consistency of abstract postulates. The procedure goes something like this. Let us understand by the word 'class' a collection or aggregate of distinguishable elements, each of which is called a member of the class. Thus, the class of prime numbers less than 10 is the collection whose members are 2, 3, 5, and 7. Suppose the following set of postulates concerning two classes K and L, whose special nature is left undetermined except as "implicitly" defined by the postulates:

1. Any two members of K are contained in just one member of L.
2. No member of K is contained in more than two members of L.
3. The members of K are not all contained in a single member of L.
4. Any two members of L contain just one member of K.
5. No member of L contains more than two members of K.

From this small set we can derive, by using customary rules of inference, a number of theorems. For example, it can be shown that K contains just three members. But is the set consistent, so that mutually contradictory theorems can never be derived from it? The question can be answered readily with the help of the following model:

> Let K be the class of points consisting of the vertices of a triangle, and L the class of lines made up of its sides; and let us understand the phrase 'a member of K is contained in a member of L' to mean that a point which is a vertex lies on a line which is a side. Each of the five abstract postulates is then converted into a true statement. For instance, the first postulate asserts that any two points which are vertices of the triangle lie on just one line which is a side. (See Fig. 1.) In this way the set of postulates is proved to be consistent.

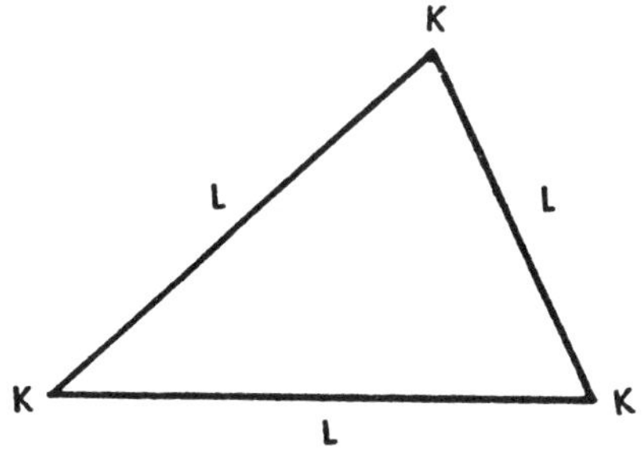

Fig. 1.
Model for a set of postulates about two classes, K and L, is a triangle whose vertices are the members of K and whose sides are the members of L. The geometrical model shows that the postulates are consistent.

The consistency of plane Riemannian geometry can also, ostensibly, be established by a model embodying the postulates. We may interpret the expression 'plane' in the Riemannian axioms to signify the surface of a Euclidean sphere, the expression 'point' a point on this surface, the expression 'straight line' an arc of a great circle on this surface, and so on. Each Riemannian postulate is then converted into a theorem of Euclid. For example, on this interpretation the Riemannian parallel postulate reads: Through a point on the surface of a sphere, no arc of a great circle can be drawn parallel to a given arc of a great circle. (See Fig. 2.)

At first glance this proof of the consistency of Riemannian geometry may seem conclusive. But a closer look is disconcerting. For a sharp eye will discern that the problem has not been solved; it has merely been shifted to another domain. The proof attempts to settle the consistency of Riemannian geometry by appealing to the consistency of Euclidean geometry. What emerges, then, is only this: Riemannian geometry is consistent if Euclidean geometry is consistent. The authority of Euclid is thus invoked to demonstrate the consistency of a system which challenges the exclusive validity of Euclid. The inescapable question is: Are the axioms of the Euclidean system itself consistent?

An answer to this question, hallowed, as we have noted, by a long tradition, is that the Euclidean axioms are true and are therefore consistent. This answer is

Fig. 2

The non-Euclidean geometry of Bernhard Riemann can be represented by a Euclidean model. The Riemannian plane becomes the surface of a Euclidean sphere, points on the plane

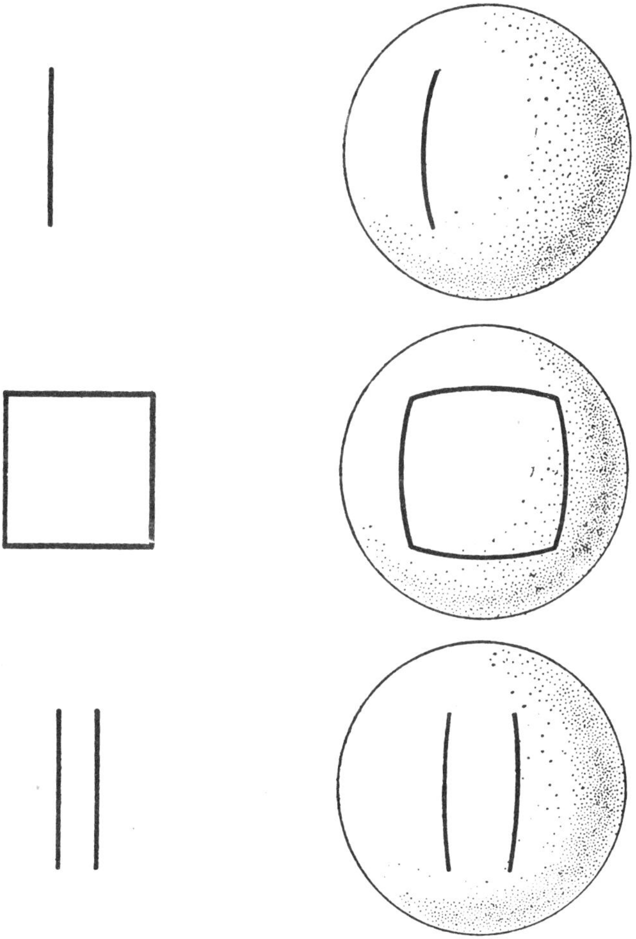

become points on this surface, straight lines in the plane become great circles. Thus, a portion of the Riemannian plane bounded by segments of straight lines is depicted as a portion of the sphere bounded by parts of great circles (*center*). Two line segments in the Riemannian plane are two segments of great circles on the Euclidean sphere (*bottom*), and these, if extended, indeed intersect, thus contradicting the parallel postulate.

no longer regarded as acceptable; we shall return to it presently and explain why it is unsatisfactory. Another answer is that the axioms jibe with our actual, though limited, experience of space and that we are justified in extrapolating from the small to the universal. But, although much inductive evidence can be adduced to support this claim, our best proof would be logically incomplete. For even if all the observed facts are in agreement with the axioms, the possibility is open that a hitherto unobserved fact may contradict them and so destroy their title to universality. Inductive considerations can show no more than that the axioms are plausible or probably true.

Hilbert tried yet another route to the top. The clue to his way lay in Cartesian coordinate geometry. In his interpretation Euclid's axioms were simply transformed into algebraic truths. For instance, in the axioms for plane geometry, construe the expression 'point' to signify a pair of numbers, the expression 'straight line' the (linear) relation between numbers expressed by a first degree equation with two unknowns, the expression 'circle' the relation between numbers expressed by a quadratic equation of a certain form, and so on. The geometric statement that two distinct points uniquely determine a straight line is then transformed into the algebraic truth that two distinct pairs of numbers uniquely determine a linear relation; the geometric theorem that a straight line intersects a circle in at most two points, into the algebraic theorem that a pair of simultaneous equations in two unknowns (one of which is linear and the other quadratic of a certain type) determine at most two pairs of real numbers; and so on. In brief, the consistency of the Euclidean postulates is established by showing that they are satisfied by an algebraic model. This method of establishing consistency is powerful and effective. Yet it, too, is vulnerable to the objection already set forth. For, again, a problem in one domain is resolved by transferring it to another. Hilbert's argument for the consistency of his geometric postulates shows that if algebra is consistent, so is his geometric system. The proof is clearly relative to the assumed consistency of another system and is not an "absolute" proof.

In the various attempts to solve the problem of consistency there is one persistent source of difficulty. It lies in the fact that the axioms are interpreted by models composed of an infinite number of elements. This makes it impossible to encompass the models in a finite number of observations; hence the truth of the axioms themselves is subject to doubt. In the inductive argument for the truth of Euclidean geometry, a finite number of observed facts about space are presumably in agreement with the axioms. But the conclusion that the argument seeks to establish involves an extrapolation from a finite to an infinite set of data. How can we justify this jump? On the other hand, the difficulty is minimized, if not completely eliminated, where an appropriate model can be devised that contains only a finite number of elements. The triangle model used to show the consistency of the five abstract postulates for the classes K and L is finite; and it is comparatively simple to determine by actual inspection whether all the elements in the model actually satisfy the postulates, and thus whether they are true (and hence consistent). To illustrate: by examining in turn all the vertices of the model triangle, one can learn whether any two of them lie on just one side—so that the first postulate is established as true. Since all the elements of the model, as well as the relevant relations among them, are open to direct and exhaustive inspection, and since the likelihood of mistakes occurring in inspecting them is practically nil, the consistency of the postulates in this case is not a matter for genuine doubt.

Unfortunately, most of the postulate systems that constitute the foundations of important branches of mathematics cannot be mirrored in finite models. Consider the postulate in elementary arithmetic which asserts that every integer has an immediate successor differing from any preceding integer. It is evident that the model needed to test the set to which this postulate belongs cannot be finite, but must contain an infinity of elements. It follows that the truth (and so the consistency) of the set cannot be established by an exhaustive inspection of a limited number of elements. Apparently we have reached an impasse. Finite models suffice, in principle, to establish the consistency of certain sets of postulates; but these are of slight mathematical importance. Non-finite models, necessary for the interpretation of most postulate systems of mathematical significance, can be described only in general terms; and we cannot conclude as a matter of course that the descriptions are free from concealed contradictions.

It is tempting to suggest at this point that we can be sure of the consistency of formulations in which non-finite models are described if the basic notions employed are transparently "clear" and "distinct." But the history of thought has not dealt kindly with the doctrine of clear and distinct ideas, or with the doctrine of intuitive knowledge implicit in the suggestion. In certain areas of mathematical research in which assumptions about infinite collections play central roles, radical contradictions have turned up, in spite of the intuitive clarity of the notions involved in the assumptions and despite the seemingly consistent character of the intellectual constructions performed. Such contradictions (technically referred to as "antinomies") have emerged in the theory of infinite numbers, developed by Georg Cantor in the nineteenth century; and the occurrence of these contradictions has made plain that the apparent clarity of even such an elementary notion as that of *class* (or *aggregate*) does not guarantee the consistency of any particular system built on it. Since the mathematical theory of classes, which deals with the properties and relations of aggregates or collections of elements, is often adopted as the foundation for other branches of mathematics, and in particular for elementary arithmetic, it is pertinent to ask whether contradictions similar to those encountered in the theory of infinite classes infect the formulations of other parts of mathematics.

In point of fact, Bertrand Russell constructed a contradiction within the framework of elementary logic itself that is precisely analogous to the contradiction first developed in the Cantorian theory of infinite classes. Russell's antinomy can be stated as follows. Classes seem to be of two kinds: those which do not contain themselves as members, and those which do. A class will be called "normal" if, and only if, it does not contain itself as a member; otherwise it will be called "non-normal." An example of a normal class is the class of mathematicians, for patently the class itself is not a mathematician and is therefore not a member of itself. An example of a non-normal class is the class of all thinkable things; for the class of all thinkable things is itself thinkable and is therefore a member of itself. Let 'N' by definition stand for the class of *all* normal classes. We ask whether N itself is a normal class. If N is normal, it is a member of itself (for by definition N contains all normal classes); but, in that case, N is non-normal, because by definition a class that contains itself as a member is non-normal. On the other hand, if N is non-normal, it is a member of itself (by definition of non-normal); but, in that case, N is normal, because by definition the members of N are normal classes. In short, N is normal if, and only if, N is non-normal. It follows that the statement 'N is normal' is both true and false. This fatal contradiction results from an uncritical use of the apparently pellucid notion of class. Other paradoxes were found later, each of them constructed by means of familiar and seemingly cogent modes of reasoning. Mathematicians came to realize that in developing consistent systems familiarity and intuitive clarity are weak reeds to lean on.

We have seen the importance of the problem of consistency, and we have acquainted ourselves with the classically standard method for solving it with the help of models. It has been shown that in most instances the problem requires the use of a non-finite model, the description of which may itself conceal inconsistencies. We must conclude that, while the model method is an invaluable mathematical tool, it does not supply a final answer to the problem it was designed to solve.

III

Absolute Proofs of Consistency

The limitations inherent in the use of models for establishing consistency, and the growing apprehension that the standard formulations of many mathematical systems might all harbor internal contradictions, led to new attacks upon the problem. An alternative to relative proofs of consistency was proposed by Hilbert. He sought to construct "absolute" proofs, by which the consistency of systems could be established without assuming the consistency of some other system. We must briefly explain this approach as a further preparation for understanding Gödel's achievement.

The first step in the construction of an absolute proof, as Hilbert conceived the matter, is the *complete formalization* of a deductive system. This involves draining the expressions occurring within the system of all meaning: they are to be regarded simply as empty signs. How these signs are to be combined and manipulated is to be set forth in a set of precisely stated rules. The purpose of this procedure is to construct a system of signs (called a "calculus") which conceals nothing and which has in it only that which we explicitly put into it. The postulates and theorems of a completely formalized system are "strings" (or finitely long sequences) of meaningless marks, constructed according to rules for combining the elementary signs of the system into larger wholes. Moreover, when a system has been completely formalized, the derivation of theorems from postulates is nothing more than the transformation (pursuant to rule) of one set of such "strings" into another set of "strings." In this way the danger is eliminated of using any unavowed principles of reasoning. Formalization is a difficult and tricky business, but it serves a valuable purpose. It reveals structure and function in naked clarity, as does a cut-away working model of a machine. When a system has been formalized, the logical relations between mathematical propositions are exposed to view; one is able to see the structural patterns of various "strings" of "meaningless" signs, how they hang together, how they are combined, how they nest in one another, and so on.

A page covered with the "meaningless" marks of such a formalized mathematics does not *assert* anything—it is simply an abstract design or a mosaic possessing a determinate structure. Yet it is clearly possible to describe the configurations of such a system and to make statements about the configurations and their various relations to one another. One may say that a "string" is pretty, or that it resembles another "string," or that one "string" appears to be made up of three others, and so on. Such statements are evidently meaningful and may convey important information about the formal system. It must now be observed, however, that such meaningful statements about a meaningless (or formalized) mathematical system plainly do not themselves belong to that system. They belong to what Hilbert called "meta-mathematics," to the language that is *about* mathematics. Meta-mathematical statements are statements about the signs occurring within a formalized mathematical system (i.e., a calculus)—about the kinds and arrangements of such signs when they are combined to form longer strings of marks called "formulas," or about the relations between formulas that may obtain as a consequence of the rules of manipulation specified for them.

A few examples will help to an understanding of Hilbert's distinction between mathematics (i.e., a system of meaningless signs) and meta-mathematics (meaningful statements about mathematics, the signs occurring in the calculus, their arrangement and relations). Consider the expression:

$$2 + 3 = 5$$

This expression belongs to mathematics (arithmetic) and is constructed entirely out of elementary arithmetical signs. On the other hand, the *statement*

'$2 + 3 = 5$' is an arithmetical formula

asserts something about the displayed expression. The statement does not express an arithmetical fact and does not belong to the formal language of arithmetic;

it belongs to meta-mathematics, because it characterizes a certain string of arithmetical signs as being a formula. The following statement belongs to meta-mathematics:

> If the sign '$=$' is to be used in a formula of arithmetic, the sign must be flanked both left and right by numerical expressions.

This statement lays down a necessary condition for using a certain arithmetical sign in arithmetical formulas: the structure that an arithmetical formula must have if it is to embody that sign.

Consider next the three formulas:

$$x = x$$
$$0 = 0$$
$$0 \neq 0$$

Each of these belongs to mathematics (arithmetic), because each is built up entirely out of arithmetical signs. But the statement:

'x' is a variable

belongs to meta-mathematics, since it characterizes a certain arithmetical sign as belonging to a specific class of signs (i.e., to the class of variables). Again, the following statement belongs to meta-mathematics:

> The formula '$0 = 0$' is derivable from the formula '$x = x$' by substituting the numeral '0' for the variable 'x'.

It specifies in what manner one arithmetical formula

can be obtained from another formula, and thereby describes how the two formulas are related to each other. Similarly, the statement

'$0 \neq 0$' is not a theorem

belongs to meta-mathematics, for it says of a certain formula that it is not derivable from the axioms of arithmetic, and thus asserts that a certain relation does not hold between the indicated formulas of the system. Finally, the next statement belongs to meta-mathematics:

Arithmetic is consistent

(i.e., it is not possible to derive from the axioms of arithmetic two formally contradictory formulas—for example, the formulas '$0 = 0$' and '$0 \neq 0$'). This is patently about arithmetic, and asserts that pairs of formulas of a certain sort do not stand in a specific relation to the formulas that constitute the axioms of arithmetic.[3]

---

[3] It is worth noting that the meta-mathematical statements given in the text do not contain as constituent parts of themselves any of the *mathematical signs and formulas* that appear in the examples. At first glance this assertion seems palpably untrue, for the signs and formulas are plainly visible. But, if the statements are examined with an analytic eye, it will be seen that the point is well taken. The meta-mathematical statements contain the *names* of certain arithmetical expressions, but not the arithmetical expressions themselves. The distinction is subtle but both valid and important. It arises out of the circumstance that the rules of English grammar require that no sentence literally contain the ob-

It may be that the reader finds the word 'metamathematics' ponderous and the concept puzzling. We shall not argue that the word is pretty; but the concept itself will perplex no one if we point out that it is used in connection with a special case of a well-known distinction, namely between a subject matter under study and discourse about the subject matter. The statement 'among phalaropes the males incubate the eggs' pertains to the subject matter investigated by zoologists, and belongs to zoology; but if we say that this assertion about phalaropes proves that zoology is irrational, our statement is not about phalaropes, but about the assertion and the discipline in which it oc-

---

jects to which the expressions in the sentence may refer, but only the *names* of such objects. Obviously, when we talk about a city we do not put the city itself into a sentence, but only the name of the city; and, similarly, if we wish to say something about a word (or other linguistic sign), it is not the word itself (or the sign) that can appear in the sentence, but only a name for the word (or sign). According to a standard convention we construct a name for a linguistic expression by placing single quotation marks around it. Our text adheres to this convention. It is correct to write:

Chicago is a populous city.

But it is incorrect to write:

Chicago is tri-syllabic.

To express what is intended by this latter sentence, one must write:

'Chicago' is tri-syllabic.

Likewise, it is incorrect to write:

$x = 5$ is an equation.

We must, instead, formulate our intent by:

'$x = 5$' is an equation.

curs, and is meta-zoology. If we say that the *id* is mightier than the *ego,* we are making noises that belong to orthodox psychoanalysis; but if we criticize this statement as meaningless and unprovable, our criticism belongs to meta-psychoanalysis. And so in the case of mathematics and meta-mathematics. The formal systems that mathematicians construct belong in the file labeled "mathematics"; the description, discussion, and theorizing about the systems belong in the file marked "meta-mathematics."

The importance to our subject of recognizing the distinction between mathematics and meta-mathematics cannot be overemphasized. Failure to respect it has produced paradoxes and confusion. Recognition of its significance has made it possible to exhibit in a clear light the logical structure of mathematical reasoning. The merit of the distinction is that it entails a careful codification of the various signs that go into the making of a formal calculus, free of concealed assumptions and irrelevant associations of meaning. Furthermore, it requires exact definitions of the operations and logical rules of mathematical construction and deduction, many of which mathematicians had applied without being explicitly aware of what they were using.

Hilbert saw to the heart of the matter, and it was upon the distinction between a formal calculus and its description that he based his attempt to build "absolute" proofs of consistency. Specifically, he sought to develop a method that would yield demonstrations of consistency as much beyond genuine logical doubt as the use of finite models for establishing the consistency of certain sets of postulates—by an analysis of a finite number of structural features of expressions in completely formalized calculi. The analysis consists in noting the various types of signs that occur in a calculus, indicating how to combine them into formulas, prescribing how formulas can be obtained from other formulas, and determining whether formulas of a given kind are derivable from others through explicitly stated rules of operation. Hilbert believed it might be possible to exhibit every mathematical calculus as a sort of "geometrical" pattern of formulas, in which the formulas stand to each other in a finite number of structural relations. He therefore hoped to show, by exhaustively examining these structural properties of expressions within a system, that formally contradictory formulas cannot be obtained from the axioms of given calculi. An essential requirement of Hilbert's program in its original conception was that demonstrations of consistency involve only such procedures as make no reference either to an infinite number of structural properties of formulas or to an infinite number of operations with formulas. Such procedures are called "finitistic"; and a proof of consistency conforming to this requirement is called "absolute." An "absolute" proof achieves its objectives by using a minimum of principles of inference, and does not assume the consistency of some other set of axioms. An absolute proof of the consistency of arithmetic, if one could be constructed, would therefore show by a finitistic meta-mathematical procedure that two contradictory formulas, such as '0 = 0' and its formal negation '~ (0 = 0)'—where the sign '~' means 'not' —cannot both be derived by stated rules of inference from the axioms (or initial formulas).[4]

It may be useful, by way of illustration, to compare meta-mathematics as a theory of proof with the theory of chess. Chess is played with 32 pieces of specified design on a square board containing 64 square subdivisions, where the pieces may be moved in accordance with fixed rules. The game can obviously be played without assigning any "interpretation" to the pieces or to their various positions on the board, although such an interpretation could be supplied if desired. For example, we could stipulate that a given pawn is to represent a certain regiment in an army, that a given square is to stand for a certain geographical region, and so on. But such stipulations (or interpretations) are not customary; and neither the pieces, nor the squares, nor the positions of the pieces on the board signify anything *outside* the game. In this sense, the pieces and their configurations on the board are "meaningless." Thus the game is analogous

---

[4] Hilbert did not give an altogether precise account of just what meta-mathematical procedures are to count as finitistic. In the original version of his program the requirements for an absolute proof of consistency were more stringent than in the subsequent explanations of the program by members of his school.

to a formalized mathematical calculus. The pieces and the squares of the board correspond to the elementary signs of the calculus; the legal positions of pieces on the board, to the formulas of the calculus; the initial positions of pieces on the board, to the axioms or initial formulas of the calculus; the subsequent positions of pieces on the board, to formulas derived from the axioms (i.e., to the theorems); and the rules of the game, to the rules of inference (or derivation) for the calculus. The parallelism continues. Although configurations of pieces on the board, like the formulas of the calculus, are "meaningless," statements about these configurations, like meta-mathematical statements about formulas, are quite meaningful. A "meta-chess" statement may assert that there are twenty possible opening moves for White, or that, given a certain configuration of pieces on the board with White to move, Black is mate in three moves. Moreover, general "meta-chess" theorems can be established whose proof involves only a finite number of permissible configurations on the board. The "meta-chess" theorem about the number of possible opening moves for White can be established in this way; and so can the "meta-chess" theorem that if White has only two Knights and the King, and Black only his King, it is impossible for White to force a mate against Black. These and other "meta-chess" theorems can, in other words, be proved by finitistic methods of reasoning, that is, by examining in turn each of a finite number of configurations that can occur under stated conditions. The aim of

Hilbert's theory of proof, similarly, was to demonstrate by such finitistic methods the impossibility of deriving certain contradictory formulas in a given mathematical calculus.

IV

The Systematic Codification of Formal Logic

There are two more bridges to cross before entering upon Gödel's proof itself. We must indicate how and why Whitehead and Russell's *Principia Mathematica* came into being; and we must give a short illustration of the formalization of a deductive system—we shall take a fragment of *Principia*—and explain how its absolute consistency can be established.

Ordinarily, even when mathematical proofs conform to accepted standards of professional rigor, they suffer from an important omission. They embody principles (or rules) of inference not explicitly formulated, of which mathematicians are frequently unaware. Take Euclid's proof that there is no greatest prime number (a number is prime if it is divisible without remainder by no number other than 1 and the number itself). The argument, cast in the form of a *reductio ad absurdum*, runs as follows:

Suppose, in contradiction to what the proof seeks to establish, that there is a greatest prime number. We designate it by 'x'. Then:

1. x is the greatest prime

2. Form the product of all primes less than or equal to x, and add 1 to the product. This yields a new number y, where $y =$

$$(2 \times 3 \times 5 \times 7 \times \ldots \times x) + 1$$

3. If y is itself a prime, then x is not the greatest prime, for y is obviously greater than x

4. If y is composite (i.e., not a prime), then again x is not the greatest prime. For if y is composite, it must have a prime divisor z; and z must be different from each of the prime numbers 2, 3, 5, 7, . . . , x, smaller than or equal to x; hence z must be a prime greater than x

5. But y is either prime or composite

6. Hence x is not the greatest prime

7. There is no greatest prime

We have stated only the main links of the proof. It can be shown, however, that in forging the complete chain a fairly large number of tacitly accepted rules of inference, as well as theorems of logic, are essential. Some of these belong to the most elementary part of formal logic, others to more advanced branches; for example, rules and theorems are incorporated that belong to the "theory of quantification." This deals with relations between statements containing such "quantifying" particles as 'all', 'some', and their synonyms. We shall exhibit one elementary theorem of logic and one rule of inference, each of which is a necessary but silent partner in the demonstration.

Look at line 5 of the proof. Where does it come from? The answer is, from the logical theorem (or necessary truth): 'Either *p* or non-*p*', where '*p*' is called a sentential variable. But how do we get line 5 from this theorem? The answer is, by using the rule of inference known as the "Rule of Substitution for Sentential Variables," according to which a statement can be derived from another containing such variables by substituting any statement (in this case, '*y* is prime') for each occurrence of a distinct variable (in this case, the variable '*p*'). The use of these rules and logical theorems is, as we have said, frequently an all but unconscious action. And the analysis that exposes them, even in such relatively simple proofs as Euclid's, depends upon advances in logical theory made only within the past one hundred years.[5] Like Molière's M. Jourdain, who spoke prose all his life without knowing it, mathematicians have been reasoning for at least two millennia without being aware of all the principles underlying what they were doing. The real nature of the tools of their craft has become evident only within recent times.

For almost two thousand years Aristotle's codification of valid forms of deduction was widely regarded as complete and as incapable of essential improvement.

[5] For a more detailed discussion of the rules of inference and logical theorems needed for obtaining lines 6 and 7 of the above proof, the reader is referred to the Appendix, no. 2.

As late as 1787, the German philosopher Immanuel Kant was able to say that since Aristotle formal logic "has not been able to advance a single step, and is to all appearances a closed and completed body of doctrine." The fact is that the traditional logic is seriously incomplete, and even fails to give an account of many principles of inference employed in quite elementary mathematical reasoning.[6] A renaissance of logical studies in modern times began with the publication in 1847 of George Boole's *The Mathematical Analysis of Logic.* The primary concern of Boole and his immediate successors was to develop an algebra of logic which would provide a precise notation for handling more general and more varied types of deduction than were covered by traditional logical principles. Suppose it is found that in a certain school those who graduate with honors are made up exactly of boys majoring in mathematics and girls not majoring in this subject. How is the class of mathematics majors made up, in terms of the other classes of students mentioned? The answer is not readily forthcoming if one uses only the apparatus of traditional logic. But with the help of Boolean algebra it can easily be shown that the class of mathematics majors consists exactly of boys graduating with honors and girls not graduating with honors.

---

[6] For example, of the principles involved in the inference: 5 is greater than 3; therefore, the square of 5 is greater than the square of 3.

All gentlemen are polite.

No bankers are polite.

No gentlemen are bankers.

$$g \subset p$$
$$b \subset \overline{p}$$
$$\therefore\ g \subset \overline{b}$$

$$g\overline{p} = 0$$
$$bp = 0$$

$$gb = 0$$

Symbolic logic was invented in the middle of the 19th century by the English mathematician George Boole. In this illustration a syllogism is translated into his notation in two different ways. In the upper group of formulas, the symbol '$\subset$' means "is contained in." Thus '$g \subset p$' says that the class of gentlemen is included in the class of polite persons. In the lower group of formulas two letters together mean the class of things having both characteristics. For example, 'bp' means the class of individuals who are bankers and polite; and the equation '$bp = 0$' says that this class has no members. A line above a letter means "not." ('$\overline{p}$', for example, means impolite.)

TABLE 1

Another line of inquiry, closely related to the work of nineteenth-century mathematicians on the foundations of analysis, became associated with the Boolean program. This new development sought to exhibit pure mathematics as a chapter of formal logic; and it received its classical embodiment in the *Principia Mathematica* of Whitehead and Russell in 1910. Mathematicians of the nineteenth century succeeded in "arithmetizing" algebra and what used to be called the "infinitesimal calculus" by showing that the various notions employed in mathematical analysis are definable exclusively in arithmetical terms (i.e., in terms of the integers and the arithmetical operations upon them). For example, instead of accepting the imaginary number $\sqrt{-1}$ as a somewhat mysterious "entity," it came to be defined as an ordered pair of integers (0, 1) upon which certain operations of "addition" and "multiplication" are performed. Similarly, the irrational number $\sqrt{2}$ was defined as a certain class of rational numbers—namely, the class of rationals whose square is less than 2. What Russell (and, before him, the German mathematician Gottlob Frege) sought to show was that *all arithmetical notions* can be defined in purely logical ideas, and that all the axioms of arithmetic can be deduced from a small number of basic propositions certifiable as purely logical truths.

To illustrate: the notion of *class* belongs to general logic. Two classes are defined as "similar" if there is a one-to-one correspondence between their members, the notion of such a correspondence being explicable in terms of other logical ideas. A class that has a single member is said to be a "unit class" (e.g., the class of satellites of the planet Earth); and the cardinal number 1 can be defined as the class of all classes similar to a unit class. Analogous definitions can be given of the other cardinal numbers; and the various arithmetical operations, such as addition and multiplication, can be defined in the notions of formal logic. An arithmetical statement, e.g., '$1 + 1 = 2$', can then be exhibited as a condensed transcription of a statement containing only expressions belonging to general logic; and such purely logical statements can be shown to be deducible from certain logical axioms.

Principia Mathematica thus appeared to advance the final solution of the problem of consistency of mathematical systems, and of arithmetic in particular, by reducing the problem to that of the consistency of formal logic itself. For, if the axioms of arithmetic are simply transcriptions of theorems in logic, the question whether the axioms are consistent is equivalent to the question whether the fundamental axioms of logic are consistent.

The Frege-Russell thesis that mathematics is only a chapter of logic has, for various reasons of detail, not won universal acceptance from mathematicians. Moreover, as we have noted, the antinomies of the Cantorian theory of transfinite numbers can be duplicated within logic itself, unless special precautions are taken to prevent this outcome. But are the measures adopted in *Principia Mathematica* to outflank the antinomies adequate to exclude *all* forms of self-contradictory constructions? This cannot be asserted as a matter of course. Therefore the Frege-Russell reduction of arithmetic to logic does not provide a final answer to the consistency problem; indeed, the problem simply emerges in a more general form. But, irrespective of the validity of the Frege-Russell thesis, two features of *Principia* have proved of inestimable value for the further study of the consistency question. *Principia* provides a remarkably comprehensive system of notation, with the help of which all statements of pure mathematics (and of arithmetic in particular) can be codified in a standard manner; and it makes explicit most of the rules of formal inference used in mathematical demonstrations (eventually, these rules were made more precise and complete). *Principia,* in sum, created the essential instrument for investigating the entire system of arithmetic as an uninterpreted calculus—that is, as a system of meaningless marks, whose formulas (or "strings") are combined and transformed in accordance with stated rules of operation.

V

An Example of a Successful Absolute Proof of Consistency

We must now attempt the second task mentioned at the outset of the preceding section, and familiarize ourselves with an important, though easily understandable, example of an absolute proof of consistency. By mastering the proof, the reader will be in a better position to appreciate the significance of Gödel's paper of 1931.

We shall outline how a small portion of *Principia,* the elementary logic of propositions, can be formalized. This entails the conversion of the fragmentary system into a calculus of uninterpreted signs. We shall then develop an absolute proof of consistency.

The formalization proceeds in four steps. First, a complete catalogue is prepared of the signs to be used in the calculus. These are its vocabulary. Second, the "Formation Rules" are laid down. They declare which of the combinations of the signs in the vocabulary are acceptable as "formulas" (in effect, as sentences). The rules may be viewed as constituting the grammar of the system. Third, the "Transformation Rules" are stated. They describe the precise structure of formulas from which other formulas of given structure are derivable. These rules are, in effect, the rules of inference. Finally, certain formulas are selected as axioms (or as "primitive formulas"). They serve as foundation for the entire system. We shall use the phrase "theorem of the system" to denote any formula that can be derived from the axioms by successively applying the Transformation Rules. By a formal "proof" (or "demonstration") we shall mean a finite sequence of formulas, each of which either is an axiom or can be derived from preceding formulas in the sequence by the Transformation Rules.[7]

For the logic of propositions (often called the sentential calculus) the vocabulary (or list of "elementary signs") is extremely simple. It consists of variables and constant signs. The variables may have sentences substituted for them and are therefore called "sentential variables." They are the letters

'*p*', '*q*', '*r*', etc.

The constant signs are either "sentential connectives" or signs of punctuation. The sentential connectives are:

'$\sim$' which is short for 'not'
(and is called the "tilde"),

---

[7] It immediately follows that axioms are to be counted among the theorems.

'V' which is short for 'or',
'⊃' which is short for 'if . . . then . . .', and
'·' which is short for 'and'.

The signs of punctuation are the left- and right-hand round parentheses, '(' and ')', respectively.

The Formation Rules are so designed that combinations of the elementary signs, which would normally have the form of sentences, are called formulas. Also, each sentential variable counts as a formula. Moreover, if the letter 'S' stands for a formula, its formal negation, namely, ~ (S), is also a formula. Similarly, if S_1 and S_2 are formulas, so are $(S_1) \vee (S_2)$, $(S_1) \supset (S_2)$, and $(S_1) \cdot (S_2)$. Each of the following is a formula: 'p', '$\sim (p)$', '$(p) \supset (q)$', '$((q) \vee (r)) \supset (p)$'. But neither '$(p)(\sim (q))$' nor '$((p) \supset (q)) \vee$' is a formula: not the first, because, while '(p)' and '$(\sim .(q))$' are both formulas, no sentential connective occurs between them; and not the second, because the connective 'V' is not, as the Rules require, flanked on both left and right by a formula.[8]

Two Transformation Rules are adopted. One of them, the *Rule of Substitution* (for sentential variables), says that from a formula containing sentential variables it is always permissible to derive another formula by uniformly substituting formulas for the varia-

---

[8] Where there is no possibility of confusion, punctuation marks (i.e., parentheses) can be dropped. Thus, instead of writing '$\sim (p)$' it is sufficient to write '$\sim p$'; and instead of '$(p) \supset (q)$', simply '$p \supset q$'.

bles. It is understood that, when substitutions are made for a variable in a formula, the same substitution must be made for *each occurrence* of the variable. For example, on the assumption that '$p \supset p$' has already been established, we can substitute for the variable 'p' the formula 'q' to get '$q \supset q$'; or we can substitute the formula '$p \vee q$' to get '$(p \vee q) \supset (p \vee q)$'. Or, if we substitute actual English sentences for 'p', we can obtain each of the following from '$p \supset p$': 'Frogs are noisy ⊃ Frogs are noisy'; '(Bats are blind ∨ Bats eat mice) ⊃ (Bats are blind ∨ Bats eat mice)'.[9] The second Transformation Rule is the *Rule of Detachment* (or *Modus Ponens*). This rule says that from two formulas having the form S_1 and $S_1 \supset S_2$ it is always permissible to derive the formula S_2. For example, from the two formulas '$p \vee \sim p$' and '$(p \vee \sim p) \supset (p \supset p)$', we can derive '$p \supset p$'.

Finally, the axioms of the calculus (essentially those of *Principia*) are the following four formulas:

1. $(p \vee p) \supset p$
or, in ordinary English, if either p or p, then p

1. If (either Henry VIII was a boor or Henry VIII was a boor) then Henry VIII was a boor

---

[9] On the other hand, suppose the formula '$(p \supset q) \supset (\sim q \supset \sim p)$' has already been established, and we decide to substitute 'r' for the variable 'p' and '$p \vee r$' for the variable 'q'. We cannot, by this substitution, obtain the formula '$(r \supset (p \vee r)) \supset (\sim q \supset \sim r)$', because we have failed to make the same substitution for *each* occurrence of the variable 'q'.

2. $p \supset (p \vee q)$
that is, if p, then either p or q

2. If psychoanalysis is fashionable, then (either psychoanalysis is fashionable or headache powders are sold cheap)

3. $(p \vee q) \supset (q \vee p)$
that is, if either p or q, then either q or p

3. If (either Immanuel Kant was punctual or Hollywood is sinful), then (either Hollywood is sinful or Immanuel Kant was punctual)

4. $(p \supset q) \supset ((r \vee p) \supset (r \vee q))$
that is, if (if p then q) then (if (either r or p) then (either r or q))

4. If (if ducks waddle then 5 is a prime) then (if (either Churchill drinks brandy or ducks waddle) then (either Churchill drinks brandy or 5 is a prime))

In the left-hand column we have stated the axioms, with a translation for each. In the right-hand column we have given an example for each axiom. The clumsiness of the translations, especially in the case of the final axiom, will perhaps help the reader to realize the advantages of using a special symbolism in formal logic. It is also important to observe that the nonsensical illustrations used as substitution instances for the axioms and the fact that the consequents bear no meaningful relation to the antecedents in the condi-

The correct substitution yields '$(r \supset (p \vee r)) \supset (\sim (p \vee r) \supset \sim r)$'.

tional sentences in no way affect the validity of the logical connections asserted in the examples.

Each of these axioms may seem "obvious" and trivial. Nevertheless, it is possible to derive from them with the help of the stated Transformation Rules an indefinitely large class of theorems which are far from obvious or trivial. For example, the formula

$$`((p \supset q) \supset ((r \supset s) \supset t)) \supset ((u \supset ((r \supset s) \supset t)) \supset ((p \supset u) \supset (s \supset t)))'$$

can be derived as a theorem. We are, however, not interested for the moment in deriving theorems from the axioms. Our aim is to show that this set of axioms is not contradictory, that is, to prove "absolutely" that it is *impossible* by using the Transformation Rules to derive from the axioms a formula S together with its formal negation $\sim$ S.

Now, it happens that '$p \supset (\sim p \supset q)$' (in words: 'if *p*, then if not-*p* then *q*') is a theorem in the calculus. (We shall accept this as a fact, without exhibiting the derivation.) Suppose, then, that some formula S as well as its contradictory $\sim$ S were deducible from the axioms. By substituting S for the variable '*p*' in the theorem (as permitted by the Rule of Substitution), and applying the Rule of Detachment twice, the formula '*q*' would be deducible.[10] But, if the formula con-

[10] By substituting S for '*p*' we first obtain: $S \supset (\sim S \supset q)$. From this, together with S, which is assumed to be demonstrable, we obtain by the Detachment Rule: $\sim S \supset q$. Finally, since $\sim$ S is also assumed to be demonstrable, using the Detachment Rule once more, we get: *q*.

sisting of the variable '*q*' is demonstrable, it follows at once that by substituting *any formula whatsoever* for '*q*', *any formula whatsoever is deducible from the axioms.* It is thus clear that, if both some formula S and its contradictory $\sim$ S were deducible from the axioms, every formula would be deducible. In short, if the calculus is not consistent, every formula is a theorem—which is the same as saying that from a contradictory set of axioms any formula can be derived. But this has a converse: namely, if not every formula is a theorem (i.e., if there is at least one formula that is not derivable from the axioms), then the calculus is consistent. *The task, therefore, is to show that there is at least one formula that cannot be derived from the axioms.*

The way this is done is to employ meta-mathematical reasoning upon the system before us. The actual procedure is elegant. It consists in finding a characteristic or structural property of formulas which satisfies the three following conditions. (1) The property must be common to all four axioms. (One such property is that of containing not more than 25 elementary signs; however, this property does not satisfy the next condition.) (2) The property must be "hereditary" under the Transformation Rules—that is, if all the axioms have the property, any formula properly derived from them by the Transformation Rules must also have it. Since any formula so derived is by definition a theorem, this condition in essence stipulates that every theorem must have the property. (3) The property must not belong to every formula that can be constructed in accordance with the Formation Rules of the system—that is, we must seek to exhibit at least one formula that does not have the property. If we succeed in this threefold task, we shall have an absolute proof of consistency. The reasoning runs something like this: the hereditary property is transmitted from the axioms to all theorems; but if an array of signs can be found that conforms to the requirements of being a formula in the system and that, nevertheless, does not possess the specified hereditary property, this formula cannot be a theorem. (To put the matter in another way, if a suspected offspring (formula) lacks an invariably inherited trait of the forebears (axioms), it cannot in fact be their descendant (theorem)). But, if a formula is discovered that is not a theorem, we have established the consistency of the system; for, as we noted a moment ago, if the system were *not* consistent, *every* formula could be derived from the axioms (i.e., every formula would be a theorem). In short, the exhibition of a single formula without the hereditary property does the trick.

Let us identify a property of the required kind. The one we choose is the property of being a "tautology." In common parlance, an utterance is usually said to be tautologous if it contains a redundancy and says the same thing twice over in different words—e.g., 'John is the father of Charles and Charles is a son of John'. In logic, however, a tautology is defined as a statement that excludes no logical possibilities—e.g., 'Either it is raining or it is not raining'. Another way of putting this is to say that a tautology is "true in all possible worlds." No one will doubt that, irrespective of the actual state of the weather (i.e., regardless of whether the statement that it is raining is true or false), the statement 'Either it is raining or it is not raining' is *necessarily true*.

We employ this notion to define a tautology in our system. Notice, first, that every formula is constructed of elementary constituents 'p', 'q', 'r', etc. A formula is a tautology if it is invariably true, regardless of whether its elementary constituents are true or false. Thus, in the first axiom '$(p \vee p) \supset p$', the only elementary constituent is 'p'; but it makes no difference whether 'p' is assumed to be true or is assumed to be false—the first axiom is true in either case. This may be made more evident if we substitute for 'p' the statement 'Mt. Rainier is 20,000 feet high'; we then obtain as an instance of the first axiom the statement 'If either Mt. Rainier is 20,000 feet high or Mt. Rainier is 20,000 feet high, then Mt. Rainier is 20,000 feet high'. The reader will have no difficulty in recognizing this long statement to be true, even if he should not happen to know whether the constituent statement 'Mt. Rainier is 20,000 feet high' is true. Obviously, then, the first axiom is a tautology—"true in all possible worlds." It can easily be shown that each of the other axioms is also a tautology.

Next, it is possible to prove that the property of being a tautology is hereditary under the Transformation Rules, though we shall not turn aside to give the demonstration. (See Appendix, no. 3.) It follows that every formula properly derived from the axioms (i.e., every theorem) must be a tautology.

It has now been shown that the property of being tautologous satisfies two of the three conditions mentioned earlier, and we are ready for the third step. We must look for a formula that belongs to the system (i.e., is constructed out of the signs mentioned in the vocabulary in accordance with the Formation Rules), yet, because it does not possess the property of being a tautology, cannot be a theorem (i.e., cannot be derived from the axioms). We do not have to look very hard; it is easy to exhibit such a formula. For example, '$p \vee q$' fits the requirements. It purports to be a gosling but is in fact a duckling; it does not belong to the family: it is a *formula,* but it is *not a theorem.* Clearly, it is not a tautology. Any substitution instance (or interpretation) shows this at once. We can obtain by substitution for the variables in '$p \vee q$' the statement 'Napoleon died of cancer or Bismarck enjoyed a cup of coffee'. This is not a truth of logic, because it would be false if both of the two clauses occurring in it were false; and, even if it is a true statement, it is not true irrespective of the truth or falsity of its constituent statements. (See Appendix, no. 3.)

We have achieved our goal. We have found at least one formula that is not a theorem. Such a formula could not occur if the axioms were contradictory. Consequently, it is not possible to derive from the axioms of the sentential calculus both a formula and its negation. In short, we have exhibited an absolute proof of the consistency of the system.[11]

Before leaving the sentential calculus, we must mention a final point. Since every theorem of this calculus is a tautology, a truth of logic, it is natural to ask whether, conversely, every logical truth expressible in the vocabulary of the calculus (i.e., every tautology) is also a theorem (i.e., derivable from the axioms). The answer is yes, though the proof is too long to be stated here. The point we are concerned with making, however, does not depend on acquaintance with the proof. The point is that, in the light of this conclusion, the axioms are sufficient for generating *all* tautologous formulas—*all* logical truths expressible in the system. Such axioms are said to be "complete."

Now, it is frequently of paramount interest to de-

---

[11] The reader may find helpful the following recapitulation of the sequence:

1. Every axiom of the system is a tautology.
2. Tautologousness is a hereditary property.
3. Every formula properly derived from the axioms (i.e., every theorem) is also a tautology.
4. Hence any formula that is not a tautology is not a theorem.
5. One formula has been found (e.g., '$p \vee q$') that is not a tautology.
6. This formula is therefore not a theorem.
7. But, if the axioms were inconsistent, every formula would be a theorem.
8. Therefore the axioms are consistent.

termine whether an axiomatized system is complete. Indeed, a powerful motive for axiomatizing various branches of mathematics has been the desire to establish a set of initial assumptions from which all the true statements in some field of inquiry are deducible. When Euclid axiomatized elementary geometry, he apparently so selected his axioms as to make it possible to derive from them all geometric truths; that is, those that had already been established, as well as any others that might be discovered in the future.[12] Until recently it was taken as a matter of course that a complete set of axioms for any given branch of mathematics can be assembled. In particular, mathematicians believed that the set proposed for arithmetic in the past was in fact complete, or, at worst, could be made complete simply by adding a finite number of axioms to the original list. The discovery that this will not work is one of Gödel's major achievements.

---

[12] Euclid showed remarkable insight in treating his famous parallel axiom as an assumption logically independent of his other axioms. For, as was subsequently proved, this axiom cannot be derived from his remaining assumptions, so that without it the set of axioms is incomplete.

VI

The Idea of Mapping and Its Use in Mathematics

The sentential calculus is an example of a mathematical system for which the objectives of Hilbert's theory of proof are fully realized. To be sure, this calculus codifies only a fragment of formal logic, and its vocabulary and formal apparatus do not suffice to develop even elementary arithmetic. Hilbert's program, however, is not so limited. It can be carried out successfully for more inclusive systems, which can be shown by meta-mathematical reasoning to be both consistent and complete. By way of example, an absolute proof of consistency is available for a system of arithmetic that permits the *addition* of cardinal numbers though not their multiplication. But is Hilbert's finitistic method powerful enough to prove the consistency of a system such as *Principia,* whose vocabulary and logical apparatus are adequate to express the whole of arithmetic and not merely a fragment? Repeated attempts to construct such a proof were unsuccessful; and the publication of Gödel's paper in 1931 showed, finally, that all such efforts operating within the strict limits of Hilbert's original program must fail.

What did Gödel establish, and how did he prove his results? His main conclusions are twofold. In the first place (though this is not the order of Gödel's actual argument), he showed that it is impossible to give a meta-mathematical proof of the consistency of a system comprehensive enough to contain the whole of arithmetic—unless the proof itself employs rules of inference in certain essential respects different from the Transformation Rules used in deriving theorems within the system. Such a proof may, to be sure, possess great value and importance. However, if the reasoning in it is based on rules of inference much more powerful than the rules of the arithmetical calculus, so that the consistency of the assumptions in the reasoning is as subject to doubt as is the consistency of arithmetic, the proof would yield only a specious victory: one dragon slain only to create another. In any event, if the proof is not finitistic, it does not realize the aims of Hilbert's original program; and Gödel's argument makes it unlikely that a finitistic proof of the consistency of arithmetic can be given.

Gödel's second main conclusion is even more surprising and revolutionary, because it demonstrates a fundamental limitation in the power of the axiomatic method. Gödel showed that *Principia,* or any other system within which arithmetic can be developed, is *essentially incomplete.* In other words, given *any* consistent set of arithmetical axioms, there are true arithmetical statements that cannot be derived from the set. This crucial point deserves illustration. Mathematics abounds in general statements to which no exceptions have been found that thus far have thwarted all attempts at proof. A classical illustration is known as "Goldbach's theorem," which states that every even number is the sum of two primes. No even number has ever been found that is not the sum of two primes, yet no one has succeeded in finding a proof that Goldbach's conjecture applies without exception to all even numbers. Here, then, is an example of an arithmetical statement that may be true, but may be non-derivable from the axioms of arithmetic. Suppose, now, that Goldbach's conjecture were indeed universally true, though not derivable from the axioms. What of the suggestion that in this eventuality the axioms could be modified or augmented so as to make hitherto unprovable statements (such as Goldbach's on our supposition) derivable in the enlarged system? Gödel's results show that even if the supposition were correct the suggestion would still provide no final cure for the difficulty. That is, even if the axioms of arithmetic are augmented by an indefinite number of other true ones, there will always be further arithmetical truths that are not formally derivable from the augmented set.[13]

---

[13] Such further truths may, as we shall see, be established by some form of meta-mathematical reasoning about an arithmetical system. But this procedure does not fit the require-

How did Gödel prove these conclusions? Up to a point the structure of his argument is modeled, as he himself pointed out, on the reasoning involved in one of the logical antinomies known as the "Richard Paradox," first propounded by the French mathematician Jules Richard in 1905. We shall outline this paradox.

Consider a language (e.g., English) in which the purely arithmetical properties of cardinal numbers can be formulated and defined. Let us examine the definitions that can be stated in the language. It is clear that, on pain of circularity or infinite regress, some terms referring to arithmetical properties cannot be defined explicitly—for we cannot define everything and must start somewhere—though they can, presumably, be understood in some other way. For our purposes it does not matter which are the undefined or "primitive" terms; we may assume, for example, that we understand what is meant by 'an integer is divisible by another', 'an integer is the product of two integers', and so on. The property of being a prime number may then be defined by: 'not divisible by any integer other than 1 and itself'; the property of being a perfect square may be defined by: 'being the product of some integer by itself'; and so on.

We can readily see that each such definition will con-

---

ment that the calculus must, so to speak, be self-contained, and that the truths in question must be exhibited as the formal consequences of the specified axioms within the system. There is, then, an *inherent* limitation in the axiomatic method as a way of systematizing the whole of arithmetic.

tain only a finite number of words, and therefore only a finite number of letters of the alphabet. This being the case, the definitions can be placed in serial order: a definition will precede another if the number of letters in the first is smaller than the number of letters in the second; and, if two definitions have the same number of letters, one of them will precede the other on the basis of the alphabetical order of the letters in each. On the basis of this order, a unique integer will correspond to each definition and will represent the number of the place that the definition occupies in the series. For example, the definition with the smallest number of letters will correspond to the number 1, the next definition in the series will correspond to 2, and so on.

Since each definition is associated with a unique integer, it may turn out in certain cases that an integer will possess the very property designated by the definition with which the integer is correlated.[14] Suppose, for instance, the defining expression 'not divisible by any integer other than 1 and itself' happens to be correlated with the order number 17; obviously 17 itself has the property designated by that expression. On the other hand, suppose the defining expression 'being the product of some integer by itself' were correlated with the order number 15; 15 clearly does not have the

[14] This is the same sort of thing that would happen if the English word 'short' appeared in a list of words, and we characterized each word of the list by the descriptive tags "short" or "long." The word 'short' would then have the tag "short" attached to it.

property designated by the expression. We shall describe the state of affairs in the second example by saying that the number 15 has the property of being *Richardian;* and, in the first example, by saying that the number 17 does *not* have the property of being *Richardian.* More generally, we define '*x* is Richardian' as a shorthand way of stating '*x* does *not* have the property designated by the defining expression with which *x* is correlated in the serially ordered set of definitions'.

We come now to a curious but characteristic turn in the statement of the Richard Paradox. The defining expression for the property of being Richardian ostensibly describes a numerical property of integers. The expression itself therefore belongs to the series of definitions proposed above. It follows that the expression is correlated with a position-fixing integer or number. Suppose this number is *n*. We now pose the question, reminiscent of Russell's antinomy: Is *n* Richardian? The reader can doubtless anticipate the fatal contradiction that now threatens. For *n* is Richardian if, and only if, *n* does not have the property designated by the defining expression with which *n* is correlated (i.e., it does not have the property of being Richardian). In short, *n* is Richardian if, and only if, *n* is not Richardian; so that the statement '*n* is Richardian' is both true and false.

We must now point out that the contradiction is, in a sense, a hoax produced by not playing the game quite fairly. An essential but tacit assumption underlying the serial ordering of definitions was conveniently dropped along the way. It was agreed to consider the definitions of the *purely arithmetical* properties of integers—properties that can be formulated with the help of such notions as arithmetical addition, multiplication, and the like. But then, without warning, we were asked to accept a definition in the series that involves reference to the *notation* used in formulating arithmetical properties. More specifically, the definition of the property of being Richardian does not belong to the series initially intended, because this definition involves meta-mathematical notions such as the number of letters (or signs) occurring in expressions. We can outflank the Richard Paradox by distinguishing carefully between statements *within* arithmetic (which make no reference to any system of notation) and statements about some system of notation in which arithmetic is codified.

The reasoning in the construction of the Richard Paradox is clearly fallacious. The construction nevertheless suggests that it may be possible to "map" or "mirror" meta-mathematical statements about a sufficiently comprehensive formal system in the system itself. The idea of "mapping" is well known and plays a fundamental role in many branches of mathematics. It is used, of course, in the construction of ordinary maps, where shapes on the surface of a sphere are projected onto a plane, so that the relations between the plane figures mirror the relations between the figures on the spherical surface. It is used in coordinate geometry, which translates geometry into algebra, so that geometric relations are mapped by algebraic ones. (The reader will recall the discussion in Section II, which explained how Hilbert used algebra to establish the consistency of his axioms for geometry. What Hilbert did, in effect, was to map geometry onto algebra.) Mapping also plays a role in mathematical physics where, for example, relations between properties of electric currents are represented in the language of hydrodynamics. And mapping is involved when a pilot model is constructed before proceeding with a full-size machine, when a small wing surface is observed for its aerodynamic properties in a wind tunnel, or when a laboratory rig made up of electric circuits is used to study the relations between large-size masses in motion. A striking visual example is presented in Fig. 3, which illustrates a species of mapping that occurs in a branch of mathematics known as projective geometry.

The basic feature of mapping is that an abstract structure of relations embodied in one domain of "objects" can be shown to hold between "objects" (usually

Fig. 3

Figure 3 (a) illustrates the Theorem of Pappus: If A, B, C are any three distinct *points* on a *line* I, and A′, B′, C′ any three distinct *points* on another *line* II, the three *points* R, S, T determined by the pair of *lines* AB′ and A′B, BC′ and B′C, CA′ and C′A, respectively, are *collinear* (i.e., lie on *line* III).

Figure 3 (b) illustrates the "dual" of the above theorem: If A, B, C are any three distinct *lines* on a *point* I, and A′, B′, C′ any three distinct *lines* on another *point* II, the three *lines* R, S, T determined by the pair of *points* AB′ and A′B, BC′ and

(a)

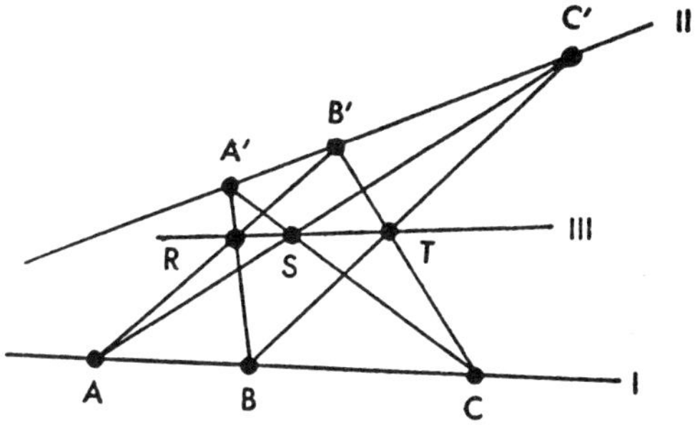

(b)

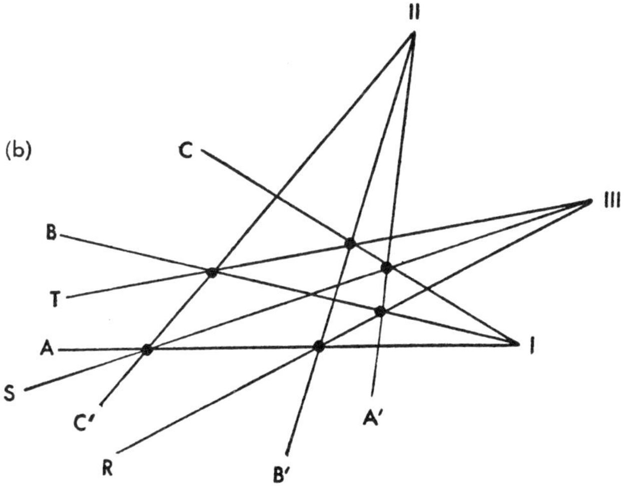

B′C, CA′ and C′A, respectively, are *copunctal* (i.e., lie on *point* III).

The two figures have the same *abstract structure,* though in appearance they are markedly different. Figure 3 (a) is so related to Figure 3 (b) that *points* of the former correspond to *lines* of the latter, while *lines* of the former correspond to *points* of the latter. In effect, (b) is a map of (a): a point in (b) represents (or is the "mirror image" of) a line in (a), while a line in (b) represents a point in (a).

of a sort different from the first set) in another domain. It is this feature which stimulated Gödel in constructing his proofs. If complicated meta-mathematical statements about a formalized system of arithmetic could, as he hoped, be translated into (or mirrored by) arithmetical statements within the system itself, an important gain would be achieved in facilitating meta-mathematical demonstrations. For just as it is easier to deal with the algebraic formulas representing (or mirroring) intricate geometrical relations between curves and surfaces in space than with the geometrical relations themselves, so it is easier to deal with the arithmetical counterparts (or "mirror images") of complex logical relations than with the logical relations themselves.

The exploitation of the notion of mapping is the key to the argument in Gödel's famous paper. Following the style of the Richard Paradox, but carefully avoiding the fallacy involved in its construction, Gödel showed that meta-mathematical statements *about* a formalized arithmetical calculus can indeed be represented by arithmetical formulas *within* the calculus. As we shall explain in greater detail in the next section, he devised a method of representation such that neither the arithmetical formula corresponding to a certain true meta-mathematical statement about the formula, nor the arithmetical formula corresponding to the denial of the statement, is demonstrable within the calculus. Since one of these arithmetical formulas must codify an arithmetical truth, yet neither is derivable from the axioms, the axioms are incomplete. Gödel's method of representation also enabled him to construct an arithmetical formula corresponding to the meta-mathematical statement 'The calculus is consistent' and to show that this formula is not demonstrable within the calculus. It follows that the meta-mathematical statement cannot be established unless rules of inference are used that cannot be represented within the calculus, so that, in proving the statement, rules must be employed whose own consistency may be as questionable as the consistency of arithmetic itself. Gödel established these major conclusions by using a remarkably ingenious form of mapping.

VII

Gödel's Proofs

Gödel's paper is difficult. Forty-six preliminary definitions, together with several important preliminary theorems, must be mastered before the main results are reached. We shall take a much easier road; nevertheless, it should afford the reader glimpses of the ascent and of the crowning structure.

A *Gödel numbering*

Gödel described a formalized calculus within which all the customary arithmetical notations can be expressed and familiar arithmetical relations established.[15] The formulas of the calculus are constructed out of a class of elementary signs, which constitute the fundamental vocabulary. A set of primitive formulas (or axioms) are the underpinning, and the theorems of the calculus are formulas derivable from the axioms with the help

---

[15] He used an adaptation of the system developed in *Principia Mathematica.* But any calculus within which the cardinal number system can be constructed would have served his purpose.

of a carefully enumerated set of Transformation Rules (or rules of inference).

Gödel first showed that it is possible to assign a *unique number* to each elementary sign, each formula (or sequence of signs), and each proof (or finite sequence of formulas). This number, which serves as a distinctive tag or label, is called the "Gödel number" of the sign, formula, or proof.[16]

The elementary signs belonging to the fundamental vocabulary are of two kinds: the constant signs and the variables. We shall assume that there are exactly ten constant signs,[17] to which the integers from 1 to 10 are attached as Gödel numbers. Most of these signs are already known to the reader: '$\sim$' (short for 'not'); '$\vee$' (short for 'or'); '$\supset$' (short for 'if . . . then . . .'); '$=$' (short for 'equals'); '0' (the numeral for the number zero); and three signs of punctuation, namely, the left parenthesis '(', the right parenthesis ')', and the comma ','. In addition, two other signs will be used: the inverted letter '$\exists$', which may be read as 'there is' and which occurs in "existential quantifiers"; and the

---

[16] There are many alternative ways of assigning Gödel numbers, and it is immaterial to the main argument which is adopted. We give a concrete example of how the numbers can be assigned to help the reader follow the discussion. The method of numbering used in the text was employed by Gödel in his 1931 paper.

[17] The number of constant signs depends on how the formal calculus is set up. Gödel in his paper used only seven constant signs. The text uses ten in order to avoid certain complexities in the exposition.

lower-case 's', which is attached to numerical expressions to designate the immediate successor of a number. To illustrate: the formula '$(\exists x)(x = s0)$' may be read 'There is an x such that x is the immediate successor of 0'. The table below displays the ten constant signs, states the Gödel number associated with each, and indicates the usual meanings of the signs.

Constant Signs	Gödel Number	Meaning
$\sim$	1	not
$\vee$	2	or
$\supset$	3	If . . . then
$\exists$	4	There is an . . .
$=$	5	equals
0	6	zero
s	7	The immediate successor of
(	8	punctuation mark
)	9	punctuation mark
,	10	punctuation mark

TABLE 2

Besides the constant elementary signs, three kinds of variables appear in the fundamental vocabulary of the calculus: the *numerical variables* 'x', 'y', 'z', etc., for which numerals and numerical expressions may be substituted; the *sentential variables* '*p*', '*q*', '*r*', etc., for which formulas (sentences) may be substituted; and the *predicate variables* '*P*', '*Q*', '*R*', etc., for which predicates, such as 'Prime' or 'Greater than', may be substituted. The variables are assigned Gödel numbers in accordance with the following rules: associate (i) with each distinct numerical variable a distinct prime number greater than 10; (ii) with each distinct sentential variable the square of a prime number greater than 10; and (iii) with each distinct predicate variable the cube of a prime greater than 10. The accompanying table illustrates the use of these rules to specify the Gödel numbers of a few variables.

Numerical Variable	Gödel Number	A Possible Substitution Instance
x	11	0
y	13	s0
z	17	y

Numerical variables are associated with prime numbers greater than 10.

Sentential Variable	Gödel Number	A Possible Substitution Instance
p	11^2	$0 = 0$
q	13^2	$(\exists x)(x = sy)$
r	17^2	$p \supset q$

Sentential variables are associated with the squares of prime numbers greater than 10.

Predicate Variable	Gödel Number	A Possible Substitution Instance
P	11^3	Prime
Q	13^3	Composite
R	17^3	Greater than

Predicate Variables are associated with the cubes of prime numbers greater than 10.

TABLE 3

Consider next a formula of the system, for example, '$(\exists x)(x = sy)$'. (Literally translated, this reads: 'There is an x such that x is the immediate successor of y', and says, in effect, that every number has an immediate successor.) The numbers associated with its ten constituent elementary signs are, respectively, 8, 4, 11, 9, 8, 11, 5, 7, 13, 9. We show this schematically below:

$$\begin{array}{cccccccccc} (& \exists & x &) & (& x & = & s & y &) \\ \downarrow & \downarrow & \downarrow & \downarrow & \downarrow & \downarrow & \downarrow & \downarrow & \downarrow & \downarrow \\ 8 & 4 & 11 & 9 & 8 & 11 & 5 & 7 & 13 & 9 \end{array}$$

It is desirable, however, to assign a single number to the formula rather than a set of numbers. This can be done easily. We agree to associate with the formula the unique number that is the product of the first ten primes in order of magnitude, each prime being raised to a power equal to the Gödel number of the corresponding elementary sign. The above formula is accordingly associated with the number

$$2^8 \times 3^4 \times 5^{11} \times 7^9 \times 11^8 \times 13^{11} \times 17^5 \times 19^7 \times 23^{13} \times 29^9;$$

let us refer to this number as m. In a similar fashion, a unique number, the product of as many primes as there are signs (each prime being raised to a power equal to the Gödel number of the corresponding sign), can be assigned to every finite sequence of elementary signs and, in particular, to every formula.[18]

Consider, finally, a sequence of formulas, such as may occur in some proof, e.g., the sequence:

$$(\exists x)(x = \mathrm{s}y)$$
$$(\exists x)(x = \mathrm{s}0)$$

The second formula when translated reads '0 has an

---

[18] Signs may occur in the calculus which do not appear in the fundamental vocabulary; these are introduced by defining them with the help of the vocabulary signs. For example, the sign '·', the sentential connective used as an abbreviation for 'and', can be defined in context as follows: '$p \cdot q$' is short for '$\sim(\sim p \vee \sim q)$'. What Gödel number is assigned to a defined sign? The answer is obvious if we notice that expressions containing defined signs can be eliminated in favor of their defining equivalents; and it is clear that a Gödel number can be determined for the transformed expressions. Accordingly, the Gödel number of the formula '$p \cdot q$' is the Gödel number of the formula '$\sim(\sim p \vee \sim q)$'. Similarly, the various numerals can be introduced by definition as follows: '1' as short for 's0', '2' as short for 'ss0', '3' as short for 'sss0', and so on. To obtain the Gödel number for the formula '$\sim(2 = 3)$', we eliminate the defined signs, thus obtaining the formula '$\sim(\mathrm{ss}0 = \mathrm{sss}0)$', and determine its Gödel number pursuant to the rules stated in the text.

immediate successor'; it is derivable from the first by substituting the numeral '0' for the numerical variable 'y'.[19] We have already determined the Gödel number of the first formula: it is m; and suppose that n is the Gödel number of the second formula. As before, it is convenient to have a single number as a tag for the sequence. We agree therefore to associate with it the number which is the product of the first two primes in order of magnitude (i.e., the primes 2 and 3), each prime being raised to a power equal to the Gödel number of the corresponding formula in the sequence. If we call this number k, we can write $k = 2^m \times 3^n$. By applying this compact procedure we can obtain a number for each sequence of formulas. In sum, every expression in the system, whether an elementary sign, a sequence of signs, or a sequence of sequences, can be assigned a unique Gödel number.

What has been done so far is to establish a method for completely "arithmetizing" the formal calculus. The method is essentially a set of directions for setting up a one-to-one correspondence between the expressions in the calculus and a certain subset of the in-

---

[19] The reader will recall that we defined a proof as a finite sequence of formulas, each of which either is an axiom or can be derived from preceding formulas in the sequence with the help of the Transformation Rules. By this definition the above sequence is not a proof, since the first formula is not an axiom and its derivation from the axioms is not shown: the sequence is only a segment of a proof. It would take too long to write out a full example of a proof, and for illustrative purposes the above sequence will suffice.

tegers.[20] Once an expression is given, the Gödel number uniquely corresponding to it can be calculated. But this is only half the story. Once a number is given, we can determine whether it is a Gödel number, and, if it is, the expression it represents can be exactly analyzed or "retrieved." If a given number is less than or equal to 10, it is the Gödel number of an elementary constant sign. The sign can be identified. If the number is greater than 10, it can be decomposed into its prime factors in just one way (as we know from a famous theorem of arithmetic).[21] If it is a prime greater than 10, or the second or third power of such a prime, it is the Gödel number of an identifiable variable. If it is the product of successive primes, each raised to some power, it may be the Gödel number either of a formula

---

[20] Not every integer is a Gödel number. Consider, for example, the number 100. 100 is greater than 10, and therefore cannot be the Gödel number of an elementary constant sign; and since it is neither a prime greater than 10, nor the square nor the cube of such a prime, it cannot be the Gödel number of a variable. On decomposing 100 into its prime factors, we find that it is equal to $2^2 \times 5^2$; and the prime number 3 does not appear as a factor in the decomposition, but is skipped. According to the rules laid down, however, the Gödel number of a formula (or of a sequence of formulas) must be the product of *successive* primes, each raised to some power. The number 100 does not satisfy this condition. In short, 100 cannot be assigned to constant signs, variables, or formulas; hence it is not a Gödel number.

[21] This theorem is known as the fundamental theorem of arithmetic. It says that if an integer is composite (i.e., not a prime) it has a unique decomposition into prime factors.

or of a sequence of formulas. In this case the expression to which it corresponds can be exactly determined. Following this program, we can take any given number apart, as if it were a machine, discover how it is constructed and what goes into it; and since each of its elements corresponds to an element of the expression it represents, we can reconstitute the expression, analyze its structure, and the like. Table 4 illustrates for a given number how we can ascertain whether it is a Gödel number and, if so, what expression it symbolizes.

A	243,000,000
B	64 × 243 × 15,625
C	$2^6 \times 3^5 \times 5^6$
D	6 5 6 ↓ ↓ ↓ 0 = 0
E	0 = 0

The arithmetical formula 'zero equals zero' has the Gödel number 243 million. Reading down from A to E, the illustration shows how the number is translated into the expression it represents; reading up, how the number for the formula is derived.

TABLE 4

B *The arithmetization of meta-mathematics*

Gödel's next step is an ingenious application of mapping. He showed that all meta-mathematical statements about the structural properties of expressions in the calculus can be adequately *mirrored* within the calculus itself. The basic idea underlying his procedure is this: Since every expression in the calculus is associated with a (Gödel) number, a meta-mathematical statement about expressions and their relations to one another may be construed as a statement about the corresponding (Gödel) numbers and their arithmetical relations to one another. In this way meta-mathematics becomes completely "arithmetized." To take a trivial analogue: customers in a busy supermarket are often given, when they enter, tickets on which are printed numbers whose order determines the order in which the customers are to be waited on at the meat counter. By inspecting the numbers it is easy to tell how many persons have been served, how many are waiting, who precedes whom, and by how many customers, and so on. If, for example, Mrs. Smith has number 37, and Mrs. Brown number 53, instead of explaining to Mrs. Brown that she has to wait her turn after Mrs. Smith, it suffices to point out that 37 is less than 53.

As in the supermarket, so in meta-mathematics. Each meta-mathematical statement is represented by a unique formula within arithmetic; and the relations of logical dependence between meta-mathematical statements are fully reflected in the numerical relations of dependence between their corresponding arithmetical formulas. Once again mapping facilitates an inquiry into structure. The exploration of meta-mathematical questions can be pursued by investigating the arithmetical properties and relations of certain integers.

We illustrate these general remarks by an elementary example. Consider the first axiom of the sentential calculus, which also happens to be an axiom in the formal system under discussion: '$(p \vee p) \supset p$'. Its Gödel number is $2^8 \times 3^{11^2} \times 5^2 \times 7^{11^2} \times 11^9 \times 13^3 \times 17^{11^2}$, which we shall designate by the letter '*a*'. Consider also the formula: '$(p \vee p)$', whose Gödel number is $2^8 \times 3^{11^2} \times 5^2 \times 7^{11^2} \times 11^9$; we shall designate it by the letter '*b*'. We now assert the meta-mathematical statement that the formula '$(p \vee p)$' is an initial part of the axiom. To what arithmetical formula in the formal system does this statement correspond? It is evident that the smaller formula '$(p \vee p)$' can be an initial part of the larger formula which is the axiom if, and only if, the (Gödel) number *b*, representing the former, is a factor of the (Gödel) number *a*, representing the latter. On the assumption that the expression 'factor of' is suitably defined in the formalized arithmetical system, the arithmetical formula which uniquely corresponds to the above meta-mathematical statement is: '*b* is a factor of *a*'. Moreover, if this formula is true, i.e., if *b* is a factor of *a*, then it is true that '$(p \vee p)$' is an initial part of '$(p \vee p) \supset p$'.

Let us fix attention on the meta-mathematical statement: 'The sequence of formulas with Gödel number *x* is a proof of the formula with Gödel number *z*'. This statement is represented (mirrored) by a definite formula in the arithmetical calculus which expresses a *purely arithmetical relation* between x and z. (We can gain some notion of the complexity of this relation by recalling the example used above, in which the Gödel number $k = 2^m \times 3^n$ was assigned to the (fragment of a) proof whose conclusion has the Gödel number n. A little reflection shows that there is here a definite, though by no means simple, arithmetical relation between k, the Gödel number of the proof, and n, the Gödel number of the conclusion.) We write this relation between x and z as the formula 'Dem (x, z)', to remind ourselves of the meta-mathematical statement to which it corresponds (i.e., of the meta-mathematical statement 'The sequence of formulas with Gödel number x is a proof (or demonstration) of the formula with Gödel number z').[22] We now ask the reader to observe that a meta-mathematical statement which says that a certain sequence of formulas is a proof for a given formula is *true*, if, and only if, the Gödel number of the alleged proof stands to the Gödel number of the conclusion in the arithmetical relation here designated by 'Dem'. Accordingly, to establish the truth or falsity of the meta-mathematical statement under discussion, we need concern ourselves only with the ques-

---

[22] The reader must keep clearly in mind that, though 'Dem (x, z)' represents the meta-mathematical statement, the formula itself belongs to the arithmetical calculus. The formula could be written in more customary notation as '$f(x, z) = 0$', where the letter 'f' denotes a complex set of arithmetical operations upon numbers. But this more customary notation does not immediately suggest the meta-mathematical interpretation of the formula.

tion whether the relation Dem holds between two numbers. Conversely, we can establish that the arithmetical relation holds between a pair of numbers by showing that the meta-mathematical statement mirrored by this relation between the numbers is true. Similarly, the meta-mathematical statement 'The sequence of formulas with the Gödel number x is *not* a proof for the formula with the Gödel number z' is represented by a definite formula in the formalized arithmetical system. This formula is the formal contradictory of 'Dem (x, z)', namely, '$\sim$ Dem (x, z)'.

One additional bit of special notation is needed for stating the crux of Gödel's argument. Begin with an example. The formula '$(\exists x)(x = sy)$' has m for its Gödel number (see page 73), while the variable 'y' has the Gödel number 13. Substitute in this formula for the variable with Gödel number 13 (i.e., for 'y') the numeral for m. The result is the formula '$(\exists x)(x = sm)$', which says literally that there is a number x such that x is the immediate successor of m. This latter formula also has a Gödel number, which can be calculated quite easily. But instead of making the calculation, we can identify the number by an unambiguous meta-mathematical characterization: it is the Gödel number of the formula that is obtained from the formula with Gödel number m, by substituting for the variable with Gödel number 13 the numeral for m. This meta-mathematical characterization uniquely determines a definite number which is a certain arithmetical function of the numbers m and 13, where the function itself can be expressed within the formalized system.[23]

---

[23] This function is quite complex. Just how complex becomes evident if we try to formulate it in greater detail. Let us attempt such a formulation, without carrying it to the bitter end. It was shown on page 73 that *m*, the Gödel number of '$(\exists x)(x = sy)$', is

$$2^8 \times 3^4 \times 5^{11} \times 7^9 \times 11^8 \times 13^{11} \times 17^5 \times 19^7 \times 23^{13} \times 29^9.$$

To find the Gödel number of '$(\exists x)(x = sm)$' (the formula obtained from the preceding one by substituting for the variable '*y*' in the latter the numeral for *m*) we proceed as follows: This formula contains the numeral '*m*', which is a *defined* sign, and, in accordance with the content of footnote 18, *m* must be replaced by its defining equivalent. When this is done, we obtain the formula:

$$(\exists x)(x = \text{ssssss} \ldots \text{s}0)$$

where the letter 's' occurs $m + 1$ times. This formula contains only the elementary signs belonging to the fundamental vocabulary, so that its Gödel number can be calculated. To do this, we first obtain the series of Gödel numbers associated with the elementary signs of the formula:

$$8, 4, 11, 9, 8, 11, 5, 7, 7, 7, \ldots 7, 6, 9$$

in which the number 7 occurs $m + 1$ times. We next take the product of the first $m + 10$ primes in order of magnitude, each prime being raised to a power equal to the Gödel number of the corresponding elementary sign. Let us refer to this number as *r*, so that

$$r = 2^8 \times 3^4 \times 5^{11} \times 7^9 \times 11^8 \times 13^{11} \times 17^5 \times 19^7 \times 23^7 \times 29^7 \times 31^7 \times \ldots \times p^9{}_{m+10}$$

where p_{m+10} is the $(m + 10)$th prime in order of magnitude.

Now compare the two Gödel numbers *m* and *r*. *m* contains a prime factor *raised to the power 13*; *r* contains all the prime factors of *m* and many others besides, but *none of them are*

The number can therefore be designated *within* the calculus. This designation will be written as 'sub (*m*, 13, *m*)', the purpose of this form being to recall the meta-mathematical characterization which it represents, viz., 'the Gödel number of the formula obtained from the formula with Gödel number *m*, by substituting for the variable with the Gödel number 13 the numeral for *m*'. We can now drop the example and generalize. The reader will see readily that the expression 'sub (*y*, 13, *y*)' is the mirror image *within* the formalized arithmetical calculus of the meta-mathematical characterization: 'the Gödel number of the formula that is obtained from the formula with Gödel number *y*, by substituting for the variable with Gödel number 13 the numeral for *y*'. He will also note that when a definite numeral is substituted for '*y*' in 'sub (*y*, 13, *y*)'—for example, the numeral for *m* or the numeral for two hundred forty three million—the resulting expression designates a definite integer which is the Gödel number of a certain formula.[24]

---

raised to the power 13. The number *r* can thus be obtained from the number *m*, by replacing the prime factor in *m* which is raised to the power 13 with other primes raised to some power different from 13. To state exactly and in full detail how *r* is related to *m* is not possible without introducing a great deal of additional notational apparatus; this is done in Gödel's original paper. But enough has been said to indicate that the number *r* is a definite arithmetical function of *m* and 13.

[24] Several questions may occur to the reader that need to be answered. It may be asked why, in the meta-mathematical

characterization just mentioned, we say that it is "the *numeral* for *y*" which is to be substituted for a certain variable, rather than "the *number y*." The answer depends on the distinction, already discussed, between mathematics and meta-mathematics, and calls for a brief elucidation of the difference between numbers and numerals. A *numeral* is a *sign*, a linguistic expression, something which one can write down, erase, copy, and so on. A *number*, on the other hand, is something which a numeral *names* or *designates*, and which cannot literally be written down, erased, copied, and so on. Thus, we say that 10 is the *number* of our fingers, and, in making this statement, we are attributing a certain "property" to the class of our fingers; but it would evidently be absurd to say that this property is a numeral. Again, the number 10 is named by the Arabic numeral '10', as well as by the Roman letter 'X'; these names are different, though they name the same number. In short, when we make a substitution for a numerical variable (which is a letter or sign) we are putting one sign in place of another sign. We cannot literally substitute a number for a sign, because a number is a property of classes (and is sometimes said to be a concept), not something we can put on paper. It follows that, in substituting for a numerical variable, we can substitute only a numeral (or some other numerical expression, such as 's0' or '7 + 5'), and not a number. This explains why, in the above meta-mathematical characterization, we state that we are substituting for the variable the *numeral* for (the number) *y*, rather than the *number y* itself.

The reader may wonder what number is designated by 'sub (*y*, 13, *y*)' if the formula whose Gödel number is *y* does not happen to contain the variable with Gödel number 13—that is, if the formula does not contain the variable '*y*'. Thus, sub (243,000,000, 13, 243,000,000) is the Gödel number of the formula obtained from the formula with Gödel number 243,000,000 by substituting for the variable '*y*' the numeral '243,000,000'. But if the reader consults Table 4, he will find that 243,000,000 is the Gödel number of the formula '0 = 0', which does not contain the variable 'y'. What, then, is the formula that is obtained from '$0 = 0$' by substituting for the variable 'y' the numeral for the number 243,000,000? The simple answer is that, since '$0 = 0$' does not contain this variable, no substitution can be made—or, what amounts to the same thing, that the formula obtained from '$0 = 0$' is this *very same* formula. Accordingly, the number designated by 'sub (243,000,000, 13, 243,000,000)' is 243,000,000.

The reader may also be puzzled as to whether 'sub (y, 13, y)' is a *formula* within the arithmetical system in the sense that, for example, '$(\exists x)(x = sy)$', '$0 = 0$', and 'Dem (x, z) ' are formulas. The answer is no, for the following reason. The expression '$0 = 0$' is called a formula, because it asserts a relation between two numbers and is thus capable of having truth or falsity significantly attributed to it. Similarly, when definite numerals are substituted for the variables in 'Dem (x, z)', this expression formulates a relation between two numbers, and so becomes a statement that is either true or false. The same holds for '$(\exists x)(x = sy)$'. On the other hand, even when a definite numeral is substituted for 'y' in 'sub (y, 13, y)', the resulting expression does not *assert* anything and therefore cannot be true or false. It merely *designates* or *names* a number, by describing it as a certain *function* of other numbers. The difference between a *formula* (which is in effect a statement about numbers, and so is either true or false) and a *name-function* (which is in effect a name that identifies a number, and so is neither true nor false) may be clarified by some further illustrations. '$5 = 3$' is a formula which, though false, declares that the two numbers 5 and 3 are equal; '$5^2 = 4^2 + 3^2$' is also a formula which asserts that a definite relation holds between the three numbers 5, 4, and 3; and, more generally, '$y = f(x)$' is a formula which asserts that a certain relation holds between the unspecified numbers x and y. On the other hand, '$2 + 3$' expresses a function of the two numbers 2 and 3, and so names a certain number (in fact, the number 5); it is not a formula, for it clearly would be nonsensical to ask whether '$2 + 3$' is true or false. '$(7 \times 5) + 8$' expresses an-

C *The heart of Gödel's argument*

At last we are equipped to follow in outline Gödel's main argument. We shall begin by enumerating the steps in a general way, so that the reader can get a bird's-eye view of the sequence.

Gödel showed (i) how to construct an arithmetical formula G that represents the meta-mathematical statement: 'The formula G is not demonstrable'. This formula G thus ostensibly says of *itself* that it is not demonstrable. Up to a point, G is constructed analogously to the Richard Paradox. In that Paradox, the expression 'Richardian' is associated with a certain number *n*, and the sentence '*n* is Richardian' is constructed. In Gödel's argument, the formula G is also associated with a certain number *h*, and is so constructed that it corresponds to the statement: 'The formula with the associated number *h* is not demonstrable'. But (ii) Gödel also showed that G is demonstrable if, and only if, its formal negation ~ G is demonstrable. This step in the argument is again analogous to a step in the Richard Paradox, in which it is proved that *n* is Richardian if, and only if, *n* is not

---

other function of the three numbers 5, 7, and 8, and designates the number 43. And, more generally, 'f(*x*)' expresses a function of *x*, and identifies a certain number when a definite numeral is substituted for '*x*' and when a definite meaning is given to the function-sign 'f'. In short, while 'Dem (*x*, *z*)' is a formula because it has the *form of a statement* about numbers, 'sub (*y*, 13, *y*)' is not a formula because it has only the *form of a name* for numbers.

Richardian. However, if a formula and its own negation are both formally demonstrable, the arithmetical calculus is not consistent. Accordingly, if the calculus is consistent, neither G nor ~ G is formally derivable from the axioms of arithmetic. Therefore, if arithmetic is consistent, G is a formally undecidable formula. Gödel then proved (iii) that, though G is not formally demonstrable, it nevertheless is a *true* arithmetical formula. It is true in the sense that it asserts that every integer possesses a certain arithmetical property, which can be exactly defined and is exhibited by whatever integer is examined. (iv) Since G is both true and formally undecidable, the axioms of arithmetic are *incomplete*. In other words, we cannot deduce all arithmetical truths from the axioms. Moreover, Gödel established that arithmetic is *essentially* incomplete: even if additional axioms were assumed so that the true formula G could be formally derived from the augmented set, another true but formally undecidable formula could be constructed. (v) Next, Gödel described how to construct an arithmetical formula A that represents the meta-mathematical statement: 'Arithmetic is consistent'; and he proved that the formula 'A ⊃ G' is formally demonstrable. Finally, he showed that the formula A is not demonstrable. From this it follows that the consistency of arithmetic cannot be established by an argument that can be represented in the formal arithmetical calculus.

Now, to give the substance of the argument more fully:

(i) The formula '∼ Dem (x, z)' has already been identified. It represents within formalized arithmetic the meta-mathematical statement: 'The sequence of formulas with the Gödel number x is not a proof for the formula with the Gödel number z'. The prefix '(x)' is now introduced into the Dem formula. This prefix performs the same function in the formalized system as does the English phrase 'For every x'. On attaching this prefix, we have a new formula: '(x) ∼ Dem (x, z)', which represents within arithmetic the meta-mathematical statement: 'For every x, the sequence of formulas with Gödel number x is not a proof for the formula with Gödel number z'. The new formula is therefore the formal paraphrase (strictly speaking, it is the unique representative), within the calculus, of the meta-mathematical statement: 'The formula with Gödel number z is not demonstrable'—or, to put it another way, 'No proof can be adduced for the formula with Gödel number z'.

What Gödel showed is that a certain special case of this formula is not formally demonstrable. To construct this special case, begin with the formula displayed as line (1):

(1) $(x) \sim \text{Dem}\,(x, \text{sub}\,(y, 13, y))$

This formula belongs to the arithmetical calculus, but it represents a meta-mathematical statement. The question is, which one? The reader should first recall that the expression 'sub $(y, 13, y)$' designates a number. This number is the Gödel number of the formula obtained from the formula with Gödel number y, by substituting for the variable with Gödel number 13 the numeral for *y*.[25] It will then be evident that the formula of line (1) represents the meta-mathematical statement: 'The formula with Gödel number sub (*y*, 13, *y*) is not demonstrable'.[26]

---

[25] It is of utmost importance to recognize that 'sub (*y*, 13, *y*)', though it is an expression in formalized arithmetic, is not a formula but rather a name-function for identifying a *number* (see explanatory footnote 24). The number so identified, however, is the Gödel number of a formula—of the formula obtained from the formula with Gödel number *y*, by substituting for the variable '*y*' the numeral for *y*.

[26] This statement can be expanded still further to read: 'The formula [whose Gödel number is the number of the formula] obtained from the formula with Gödel number *y*, by substituting for the variable with Gödel number 13 the numeral for *y*, is not demonstrable'.

The reader may be puzzled by the fact that, in the meta-mathematical statement 'The formula with Gödel number sub (*y*, 13, *y*) is not demonstrable', the expression 'sub (*y*, 13, *y*)' does not appear within quotation marks, although it has been repeatedly stated in the text that 'sub (*y*, 13, *y*)' is an *expression*. The point involved hinges once more on the distinction between using an expression to talk about what the expression designates (in which case the expression is not placed within quotation marks) and talking about the expression itself (in which case we must use a name for the expression and, in conformity with the convention for constructing such names, must place the expression within quotation marks). An example will help. '7 + 5' is an expression which designates a number; on the other hand, 7 + 5 is a number, and not an expression. Similarly, 'sub (243,000,000, 13, 243,000,000)' is an expression which designates the Gödel number of a formula (see Table 4); but sub (243,000,000, 13, 243,000,000) is the Gödel number of a formula, and is not an expression.

But, since the formula of line (1) belongs to the arithmetical calculus, it has a Gödel number that can actually be calculated. Suppose the number to be n. We now substitute for the variable with Gödel number 13 (i.e., for the variable 'y') in the formula of line (1) the numeral for n. A new formula is then obtained, which we shall call 'G' (after Gödel) and display under that label:

(G) $(x) \sim \text{Dem}\ (x, \text{sub}\ (n, 13, n))$

Formula G is the special case we promised to construct.

Now, this formula occurs within the arithmetical calculus, and therefore must have a Gödel number. What is the number? A little reflection shows that it is sub $(n, 13, n)$. To grasp this, we must recall that sub $(n, 13, n)$ is the Gödel number of the formula that is obtained from the formula with Gödel number n by substituting for the variable with Gödel number 13 (i.e., for the variable 'y') the numeral for n. But the formula G has been obtained from the formula with Gödel number n (i.e., from the formula displayed on line (1)) by substituting for the variable 'y' occurring in it the numeral for n. Hence the Gödel number of G is in fact sub $(n, 13, n)$.

But we must also remember that the formula G is the mirror image *within* the arithmetical calculus of the meta-mathematical statement: 'The formula with Gödel number sub $(n, 13, n)$ is not demonstrable'. It follows that the *arithmetical formula* '$(x) \sim \text{Dem}\ (x,$

sub $(n, 13, n)$)' *represents* in the calculus the *meta-mathematical statement:* 'The formula '$(x) \sim$ Dem $(x,$ sub $(n, 13, n)$)' is not demonstrable'. In a sense, therefore, this arithmetical formula G can be construed as asserting of itself that it is not demonstrable.

(ii) We come to the next step, the proof that G is not formally demonstrable. Gödel's demonstration resembles the development of the Richard Paradox, but stays clear of its fallacious reasoning.[27] The argument is relatively unencumbered. It proceeds by showing that *if* the formula G were demonstrable then its formal

---

[27] It may be useful to make explicit the resemblance as well as the dissimilarity of the present argument to that used in the Richard Paradox. The main point to observe is that the formula G is not identical with the meta-mathematical statement with which it is associated, but only *represents* (or mirrors) the latter within the arithmetical calculus. In the Richard Paradox (as explained on p. 63 above) the number n is the number associated with a certain *meta-mathematical* expression. In the Gödel construction, the number n is associated with a certain *arithmetical formula* belonging to the formal calculus, though this arithmetical formula in fact represents a meta-mathematical statement. (The formula represents this statement, because the meta-mathematics of arithmetic has been mapped onto arithmetic.) In developing the Richard Paradox, the question is asked whether the number n possesses the *meta-mathematical* property of being Richardian. In the Gödel construction, the question asked is whether the number sub $(n, 13, n)$ possesses a certain *arithmetical* property—namely, the arithmetical property expressed by the formula '$(x) \sim$ Dem (x, z)'. There is therefore no confusion in the Gödel construction between statements *within* arithmetic and statements *about* arithmetic, such as occurs in the Richard Paradox.

contradictory (namely, the formula '$\sim (x) \sim$ Dem $(x$, sub $(n, 13, n))$') would also be demonstrable; and, conversely, that *if* the formal contradictory of G were demonstrable then G itself would also be demonstrable. Thus we have: G is demonstrable if, and only if, $\sim$ G is demonstrable.[28] But as we noted earlier, if a

[28] This is not what Gödel actually proved; and the statement in the text, an adaptation of a theorem obtained by J. Barkley Rosser in 1936, is used for the sake of simplicity in exposition. What Gödel actually showed is that if G is demonstrable then $\sim$ G is demonstrable (so that arithmetic is then inconsistent); and if $\sim$ G is demonstrable then arithmetic is ω-inconsistent. What is ω-inconsistency? Let 'P' be some arithmetical predicate. Then arithmetic would be ω-inconsistent if it were possible to demonstrate both the formula '$(\exists x)P(x)$' (i.e., 'There is at least one number that has the property P') and also each of the infinite set of formulas '$\sim P(0)$', '$\sim P(1)$', '$\sim P(2)$', etc. (i.e., '0 does not have the property P', '1 does not have the property P', '2 does not have the property P', and so on). A little reflection shows that if a calculus is inconsistent then it is also ω-inconsistent; but the converse does not necessarily hold: a system may be ω-inconsistent without being inconsistent. For a system to be inconsistent, both '$(\exists x)P(x)$' and '$(x) \sim P(x)$' must be demonstrable. However, although if a system is ω-inconsistent both '$(\exists x) P(x)$' and each of the infinite set of formulas '$\sim P(0)$', '$\sim P(1)$', '$\sim P(2)$', etc., are demonstrable, the formula '$(x) \sim P(x)$' may nevertheless not be demonstrable, so that the system is not inconsistent.

We outline the first part of Gödel's argument that if G is demonstrable then $\sim$ G is demonstrable. Suppose the formula G were demonstrable. Then there must be a sequence of formulas within arithmetic that constitutes a proof for G. Let the Gödel number of this proof be k. Accordingly, the arithmetical relation designated by 'Dem (x, z)' must hold between formula and its formal negation can both be derived from a set of axioms, the axioms are not consistent. Whence, if the axioms of the formalized system of arithmetic are consistent, neither the formula G nor its negation is demonstrable. In short, if the axioms are consistent, G is formally *undecidable*—in the precise technical sense that neither G nor its contradictory can be formally deduced from the axioms.

(iii) This conclusion may not appear at first sight to be of capital importance. Why is it so remarkable, it may be asked, that a formula can be constructed within arithmetic which is undecidable? There is a surprise in store which illuminates the profound implications of this result. For, although the formula G is undecidable if the axioms of the system are consistent, it

---

k, the Gödel number of the proof, and sub (*n*, 13, *n*), the Gödel number of G, which is to say that 'Dem (*k*, sub (*n*, 13, *n*))' must be a true arithmetical formula. However, it can be shown that this arithmetical relation is of such type that, if it holds between a definite pair of numbers, the formula that expresses this fact is demonstrable. Consequently, the formula 'Dem (*k*, sub (*n*, 13, *n*))' is not only true, but also formally demonstrable; that is, the formula is a *theorem*. But, with the help of the Transformation Rules in elementary logic, we can immediately derive from this theorem the formula '∼ (x) ∼ Dem (x, sub (*n*, 13, *n*))'. We have therefore shown that if the formula G is demonstrable its formal negation is demonstrable. It follows that if the formal system is consistent the formula G is not demonstrable.

A somewhat analogous but more complicated argument is required to show that if ∼ G is demonstrable then G is also demonstrable. We shall not attempt to outline it.

can nevertheless be shown by *meta-mathematical* reasoning that G *is true.* That is, it can be shown that G formulates a complex but definite numerical property which necessarily holds of all integers—just as the formula '$(x) \sim (x + 3 = 2)$' (which, when it is interpreted in the usual way, says that no cardinal number, when added to 3, yields a sum equal to 2) expresses another, likewise necessary (though much simpler) property of all integers. The reasoning that validates the truth of the undecidable formula G is straightforward. First, on the assumption that arithmetic is consistent, the meta-mathematical statement 'The formula '$(x) \sim$ Dem $(x$, sub $(n, 13, n))$' is not demonstrable' has been proven true. Second, this statement is represented within arithmetic by the very formula mentioned in the statement. Third, we recall that meta-mathematical statements have been mapped onto the arithmetical formalism in such a way that true meta-mathematical statements correspond to true arithmetical formulas. (Indeed, the setting up of such a correspondence is the *raison d'être* of the mapping; as, for example, in analytic geometry where, by virtue of this process, true geometric statements always correspond to true algebraic statements.) It follows that the formula G, which corresponds to a true meta-mathematical statement, must be true. It should be noted, however, that we have established an arithmetical truth, not by deducing it formally from the axioms of arithmetic, but by a meta-mathematical argument.

(iv) We now remind the reader of the notion of

"completeness" introduced in the discussion of the sentential calculus. It was explained that the axioms of a deductive system are "complete" if every true statement that can be expressed in the system is formally deducible from the axioms. If this is not the case, that is, if not every true statement expressible in the system is deducible, the axioms are "incomplete." But, since we have just established that G is a true formula of arithmetic not formally deducible within it, it follows that the axioms of arithmetic are incomplete—on the hypothesis, of course, that they are consistent. Moreover, they are *essentially* incomplete: even if G were added as a further axiom, the augmented set would still not suffice to yield formally *all* arithmetical truths. For, if the initial axioms were augmented in the suggested manner, another true but undecidable arithmetical formula could be constructed in the enlarged system; such a formula could be constructed merely by repeating in the new system the procedure used originally for specifying a true but undecidable formula in the initial system. This remarkable conclusion holds, no matter how often the initial system is enlarged. We are thus compelled to recognize a fundamental limitation in the power of the axiomatic method. Against previous assumptions, the vast continent of arithmetical truth cannot be brought into systematic order by laying down once for all a set of axioms from which *every* true arithmetical statement can be formally derived.

(v) We come to the coda of Gödel's amazing intellectual symphony. The steps have been traced by which he grounded the meta-mathematical statement: 'If arithmetic is consistent, it is incomplete'. But it can also be shown that this conditional statement *taken as a whole* is represented by a *demonstrable* formula within formalized arithmetic.

This crucial formula can be easily constructed. As we explained in Section V, the meta-mathematical statement 'Arithmetic is consistent' is equivalent to the statement 'There is at least one formula of arithmetic that is not demonstrable'. The latter is represented in the formal calculus by the following formula, which we shall call 'A':

$$(\text{A}) \qquad (\exists y)(x) \sim \text{Dem}\,(x, y)$$

In words, this says: 'There is at least one number y such that, for every number x, x does not stand in the relation Dem to y'. Interpreted meta-mathematically, the formula asserts: 'There is at least one formula of arithmetic for which no sequence of formulas constitutes a proof'. The formula A therefore represents the antecedent clause of the meta-mathematical statement 'If arithmetic is consistent, it is incomplete'. On the other hand, the consequent clause in this statement—namely, 'It [arithmetic] is incomplete'—follows directly from 'There is a true arithmetical statement that is not formally demonstrable in arithmetic'; and the latter, as the reader will recognize, is represented in the arithmetical calculus by an old friend, the formula G. Accordingly, the conditional meta-mathematical statement 'If arithmetic is consistent, it is incomplete' is represented by the formula:

$$(\exists y)(x) \sim \text{Dem}\,(x, y) \supset (x) \sim \text{Dem}\,(x, \text{sub}\,(n, 13, n))$$

which, for the sake of brevity, can be symbolized by '$A \supset G$'. (This formula can be proved formally demonstrable, but we shall not in these pages undertake the task.)

We now show that the formula A is not demonstrable. For suppose it were. Then, since $A \supset G$ is demonstrable, by use of the Rule of Detachment the formula G would be demonstrable. But, unless the calculus is inconsistent, G is formally undecidable, that is, not demonstrable. Thus if arithmetic is consistent, the formula A is not demonstrable.

What does this signify? The formula A represents the meta-mathematical statement 'Arithmetic is consistent'. If, therefore, this statement could be established by any argument that can be mapped onto a sequence of formulas which constitutes a proof in the arithmetical calculus, the formula A would itself be demonstrable. But this, as we have just seen, is impossible, if arithmetic is consistent. The grand final step is before us: we must conclude that if arithmetic is consistent its consistency cannot be established by any meta-mathematical reasoning that can be represented within the formalism of arithmetic!

This imposing result of Gödel's analysis should not be misunderstood: it does *not* exclude a meta-mathematical proof of the consistency of arithmetic. What it excludes is a proof of consistency that can be mirrored by the formal deductions of arithmetic.[29] Meta-mathematical proofs of the consistency of arithmetic have, in fact, been constructed, notably by Gerhard Gentzen, a member of the Hilbert school, in 1936, and by others since then.[30] These proofs are of great logical significance, among other reasons because they propose new forms of meta-mathematical constructions, and because they thereby help make clear how the class of rules of inference needs to be enlarged if the consistency of arithmetic is to be established. But these proofs cannot be represented within the arithmetical calculus; and, since they are not finitistic, they do not achieve the proclaimed objectives of Hilbert's original program.

---

[29] The reader may be helped on this point by the reminder that, similarly, the proof that it is impossible to trisect an arbitrary angle with compass and straight-edge does *not* mean that an angle cannot be trisected by any means whatever. On the contrary, an arbitrary angle can be trisected if, for example, in addition to the use of compass and straight-edge, one is permitted to employ a fixed distance marked on the straight-edge.

[30] Gentzen's proof depends on arranging all the demonstrations of arithmetic in a linear order according to their degree of "simplicity." The arrangement turns out to have a pattern that is of a certain "transfinite ordinal" type. (The theory of transfinite ordinal numbers was created by the German mathematician Georg Cantor in the nineteenth century.) The proof of consistency is obtained by applying to this linear order a rule of inference called "the principle of transfinite induction." Gentzen's argument cannot be mapped onto the formalism of arithmetic. Moreover, although most students do not question the cogency of the proof, it is not finitistic in the sense of Hilbert's original stipulations for an absolute proof of consistency.

VIII

Concluding Reflections

The import of Gödel's conclusions is far-reaching, though it has not yet been fully fathomed. These conclusions show that the prospect of finding for every deductive system (and, in particular, for a system in which the whole of arithmetic can be expressed) an absolute proof of consistency that satisfies the finitistic requirements of Hilbert's proposal, though not logically impossible, is most unlikely.[31] They show also that there is an endless number of true arithmetical statements which cannot be formally deduced from any given set of axioms by a closed set of rules of inference. It follows that an axiomatic approach to number the-

[31] The possibility of constructing a finitistic absolute proof of consistency for arithmetic is not excluded by Gödel's results. Gödel showed that no such proof is possible that can be represented within arithmetic. His argument does not eliminate the possibility of strictly finitistic proofs that cannot be represented within arithmetic. But no one today appears to have a clear idea of what a finitistic proof would be like that is *not* capable of formulation within arithmetic.

ory, for example, cannot exhaust the domain of arithmetical truth. It follows, also, that what we understand by the process of mathematical proof does not coincide with the exploitation of a formalized axiomatic method. A formalized axiomatic procedure is based on an initially determined and fixed set of axioms and transformation rules. As Gödel's own arguments show, no antecedent limits can be placed on the inventiveness of mathematicians in devising new rules of proof. Consequently, no final account can be given of the precise logical form of valid mathematical demonstrations. In the light of these circumstances, whether an all-inclusive definition of mathematical or logical truth can be devised, and whether, as Gödel himself appears to believe, only a thoroughgoing philosophical "realism" of the ancient Platonic type can supply an adequate definition, are problems still under debate and too difficult for further consideration here.[32]

---

[32] Platonic realism takes the view that mathematics does not create or invent its "objects," but discovers them as Columbus discovered America. Now, if this is true, the objects must in some sense "exist" prior to their discovery. According to Platonic doctrine, the objects of mathematical study are not found in the spatio-temporal order. They are disembodied eternal Forms or Archetypes, which dwell in a distinctive realm accessible only to the intellect. On this view, the triangular or circular shapes of physical bodies that can be perceived by the senses are not the proper objects of mathematics. These shapes are merely imperfect embodiments of an indivisible "perfect" Triangle or "perfect" Circle, which is uncreated, is never fully manifested by material things, and can

Gödel's conclusions bear on the question whether a calculating machine can be constructed that would match the human brain in mathematical intelligence. Today's calculating machines have a fixed set of directives built into them; these directives correspond to the fixed rules of inference of formalized axiomatic procedure. The machines thus supply answers to problems by operating in a step-by-step manner, each step being controlled by the built-in directives. But, as Gödel showed in his incompleteness theorem, there are innumerable problems in elementary number theory that fall outside the scope of a fixed axiomatic method, and that such engines are incapable of answering, however intricate and ingenious their built-in mechanisms may be and however rapid their operations. Given a definite problem, a machine of this type might be built for solving it; but no one such machine can be built for solving every problem. The human brain may, to be sure, have built-in limitations of its own, and there may be mathematical problems it is incapable of solving. But, even so, the brain appears to embody a structure of rules of operation which is far more powerful

---

be grasped solely by the exploring mind of the mathematician. Gödel appears to hold a similar view when he says, "Classes and concepts may . . . be conceived as real objects . . . existing independently of our definitions and constructions. It seems to me that the assumption of such objects is quite as legitimate as the assumption of physical bodies and there is quite as much reason to believe in their existence" (Kurt Gödel, "Russell's Mathematical Logic," in *The Philosophy of Bertrand Russell* (ed. Paul A. Schilpp, Evanston and Chicago, 1944), p. 137).

than the structure of currently conceived artificial machines. There is no immediate prospect of replacing the human mind by robots.

Gödel's proof should not be construed as an invitation to despair or as an excuse for mystery-mongering. The discovery that there are arithmetical truths which cannot be demonstrated formally does not mean that there are truths which are forever incapable of becoming known, or that a "mystic" intuition (radically different in kind and authority from what is generally operative in intellectual advances) must replace cogent proof. It does not mean, as a recent writer claims, that there are "ineluctable limits to human reason." It does mean that the resources of the human intellect have not been, and cannot be, fully formalized, and that new principles of demonstration forever await invention and discovery. We have seen that mathematical propositions which cannot be established by formal deduction from a given set of axioms may, nevertheless, be established by "informal" meta-mathematical reasoning. It would be irresponsible to claim that these formally indemonstrable truths established by meta-mathematical arguments are based on nothing better than bare appeals to intuition.

Nor do the inherent limitations of calculating machines imply that we cannot hope to explain living matter and human reason in physical and chemical terms. The possibility of such explanations is neither precluded nor affirmed by Gödel's incompleteness theorem. The theorem does indicate that the structure and power of the human mind are far more complex and subtle than any non-living machine yet envisaged. Gödel's own work is a remarkable example of such complexity and subtlety. It is an occasion, not for dejection, but for a renewed appreciation of the powers of creative reason.

Appendix

Notes

1. (page 11) It was not until 1899 that the arithmetic of cardinal numbers was axiomatized, by the Italian mathematician Giuseppe Peano. His axioms are five in number. They are formulated with the help of three undefined terms, acquaintance with the latter being assumed. The terms are: '*number*', '*zero*', and '*immediate successor of*'. Peano's axioms can be stated as follows:

> 1. Zero is a number.
> 2. The immediate successor of a number is a number.
> 3. Zero is not the immediate successor of a number.
> 4. No two numbers have the same immediate successor.
> 5. Any property belonging to zero, and also to the immediate successor of every number that has the property, belongs to all numbers.

The last axiom formulates what is often called the "principle of mathematical induction."

2. (page 39) The reader may be interested in seeing a fuller account than the text provides of the logical theorems and rules of inference tacitly employed even in elementary mathematical demonstrations. We shall first analyze the reasoning that yields line 6 in Euclid's proof, from lines 3, 4, and 5.

We designate the letters 'p', 'q', and 'r' as "sentential variables," because sentences may be substituted for them. Also, to economize space, we write conditional statements of the form 'if p then q' as '$p \supset q$'; and we call the expression to the left of the horseshoe sign '$\supset$' the "antecedent," and the expression to the right of it the "consequent." Similarly, we write '$p \vee q$' as short for the alternative form 'either p or q'.

There is a theorem in elementary logic which reads:

$$(p \supset r) \supset [(q \supset r) \supset ((p \vee q) \supset r)]$$

It can be shown that this formulates a *necessary truth*. The reader will recognize that this formula states more compactly what is conveyed by the following much longer statement:

If (if p then r), then [if (if q then r) then (if (either p or q) then r)]

As pointed out in the text, there is a rule of inference in logic called the Rule of Substitution for Sentential Variables. According to this Rule, a sentence S_2 follows logically from a sentence S_1 which contains sentential variables, if the former is obtained from the latter by uniformly substituting any sentences for the variables. If we apply this rule to the theorem just mentioned, substituting '*y* is prime' for '*p*', '*y* is composite' for '*q*', and '*x* is not the greatest prime' for '*r*', we obtain the following:

$$(y \text{ is prime} \supset x \text{ is not the greatest prime})$$
$$\supset [(y \text{ is composite} \supset x \text{ is not the greatest prime})$$
$$\supset ((y \text{ is prime} \vee y \text{ is composite}) \supset x \text{ is not the greatest prime})]$$

The reader will readily note that the conditional sentence within the first pair of parentheses (it occurs on the first line of this instance of the theorem) simply duplicates line 3 of Euclid's proof. Similarly, the conditional sentence within the first pair of parentheses inside the square brackets (it occurs as the second line of the instance of the theorem) duplicates line 4 of the proof. Also, the alternative sentence inside the square brackets duplicates line 5 of the proof.

We now make use of another rule of inference known as the Rule of Detachment (or "Modus Ponens"). This rule permits us to infer a sentence S_2 from two other sentences, one of which is S_1 and the other, $S_1 \supset S_2$. We apply this Rule three times: first, using line 3 of Euclid's proof and the above instance of the logical theorem; next, the result obtained by this application and line 4 of the proof; and, finally, this latest result of the application and line 5 of the proof. The outcome is line 6 of the proof.

The derivation of line 6 from lines 3, 4, and 5 thus involves the tacit use of two rules of inference and a theorem of logic. The theorem and rules belong to the elementary part of logical theory, the sentential calculus. This deals with the logical relations between statements compounded out of other statements with the help of sentential connectives, of which '$\supset$' and '$\vee$' are examples. Another such connective is the conjunction 'and', for which the dot '$\cdot$' is used as a shorthand form: thus the conjunctive statement '*p* and *q*' is written as '$p \cdot q$'. The sign '$\sim$' represents the negative particle 'not'; thus 'not-*p*' is written as '$\sim p$'.

Let us examine the transition in Euclid's proof from line 6 to line 7. This step cannot be analyzed with the help of the sentential calculus alone. A rule of inference is required which belongs to a more advanced part of logical theory—namely, that which takes note of the internal complexity of statements embodying expressions such as 'all', 'every', 'some', and their synonyms. These are traditionally called *quantifiers,* and the branch of logical theory that discusses their role is the theory of quantification.

It is necessary to explain some of the notation employed in this more advanced sector of logic, as a preliminary to analyzing the transition in question. In addition to the sentential variables for which sentences may be substituted, we must consider the category of "individual variables," such as '*x*', '*y*', '*z*', etc., for which the names of individuals can be substituted. Using these variables, the universal statement 'All primes greater than 2 are odd' can be rendered: 'For every *x*, if *x* is a prime greater than 2, then *x* is odd'. The expression 'for every *x*' is called the *universal quantifier,* and in current logical notation is abbreviated by the sign '(x)'. The universal statement may therefore be written:

$$(x)(x \text{ is a prime greater than } 2 \supset x \text{ is odd})$$

Furthermore, the "particular" (or "existential") statement 'Some integers are composite' can be rendered by 'There is at least one x such that x is an integer and x is composite'. The expression 'there is at least one x' is called the *existential quantifier,* and is currently abbreviated by the notation '$(\exists x)$'. The existential statement just mentioned can be transcribed:

$$(\exists x)(x \text{ is an integer} \cdot x \text{ is composite})$$

It is now to be observed that many statements implicitly use more than one quantifier, so that in exhibiting their true structure several quantifiers must appear. Before illustrating this point, let us adopt certain abbreviations for what are usually called predicate expressions or, more simply, predicates. We shall use '$\mathrm{Pr}\ (x)$' as short for 'x is a prime number'; and '$\mathrm{Gr}\ (x, z)$' as short for 'x is greater than z'. Consider the statement: 'x is the greatest prime'. Its meaning can be made more explicit by the following locution: 'x is a prime, and, for every z which is a prime but different from x, x is greater than z'. With the help of our various abbreviations, the statement 'x is the greatest prime' can be written:

$$\mathrm{Pr}\ (x) \cdot (z)\ [(\mathrm{Pr}\ (z) \cdot \sim (x = z)) \supset \mathrm{Gr}\ (x, z)]$$

Literally, this says: 'x is a prime and, for every z, if z is a prime and z is not equal to x then x is greater than z'. We recognize in the symbolic sequence a formal, painfully explicit rendition of the content of line 1 in Euclid's proof.

Next, consider how to express in our notation the statement 'x is not the greatest prime', which appears as line 6 of the proof. This can be presented as:

$$\mathrm{Pr}\,(x) \cdot (\exists z)\,[\mathrm{Pr}\,(z) \cdot \mathrm{Gr}\,(z, x)]$$

Literally, it says: 'x is a prime and there is at least one z such that z is a prime and z is greater than x'.

Finally, the conclusion of Euclid's proof, line 7, which asserts that there is no greatest prime, is symbolically transcribed by:

$$(x)\ [\mathrm{Pr}\,(x) \supset (\exists z)(\mathrm{Pr}\,(z) \cdot \mathrm{Gr}\,(z, x))]$$

which says: 'For every x, if x is a prime, there is at least one z such that z is a prime and z is greater than x'. The reader will observe that Euclid's conclusion implicitly involves the use of more than one quantifier.

We are ready to discuss the step from Euclid's line 6 to line 7. There is a theorem in logic which reads:

$$(p \cdot q) \supset (p \supset q)$$

or when translated, 'If both p and q, then (if p then q)'. Using the Rule of Substitution, and substituting '$\mathrm{Pr}\,(x)$' for 'p', and '$(\exists z)\,[\mathrm{Pr}\,(z) \cdot \mathrm{Gr}\,(z, x)]$' for '$q$', we obtain:

$$(\mathrm{Pr}\,(x) \cdot (\exists z)\,[\mathrm{Pr}\,(z) \cdot \mathrm{Gr}\,(z, x)]) \supset$$
$$(\mathrm{Pr}\,(x) \supset (\exists z)\,[\mathrm{Pr}\,(z) \cdot \mathrm{Gr}\,(z, x)])$$

The antecedent (first line) of this instance of the

theorem simply duplicates line 6 of Euclid's proof; if we apply the Rule of Detachment, we get

$$(\mathrm{Pr}\,(x) \supset (\exists z)\ [\mathrm{Pr}\,(z) \cdot \mathrm{Gr}\,(z,\ x)])$$

According to a Rule of Inference in the logical theory of quantification, a sentence S_2 having the form '$(x)(\ldots x \ldots)$' can always be inferred from a sentence S_1 having the form '$(\ldots x \ldots)$'. In other words, the sentence having the quantifier '(x)' as a prefix can be derived from the sentence that does not contain the prefix but is like the former in other respects. Applying this rule to the sentence last displayed, we have line 7 of Euclid's proof.

The moral of our story is that the proof of Euclid's theorem tacitly involves the use not only of theorems and rules of inference belonging to the sentential calculus, but also of a rule of inference in the theory of quantification.

3. (page 54) The careful reader may demur at this point. His reservations may run something like this. The property of being a tautology has been defined in notions of truth and falsity. Yet these notions obviously involve a reference to something *outside* the formal calculus. Therefore, the procedure mentioned in the text in effect offers an *interpretation* of the calculus, by supplying a model for the system. This being so, the authors have not done what they promised, namely, to define a property of formulas in terms of purely structural features of the formulas themselves.

It seems that the difficulty noted in Section II of the text—that proofs of consistency which are based on models, and which argue from the truth of axioms to their consistency, merely shift the problem—has not, after all, been successfully outflanked. Why then call the proof "absolute" rather than relative?

The objection is well taken when directed against the exposition in the text. But we adopted this form so as not to overwhelm the reader unaccustomed to a highly abstract presentation resting on an intuitively opaque proof. Because more venturesome readers may wish to be exposed to the real thing, to see an unprettified definition that is not open to the criticisms in question, we shall supply it.

Remember that a formula of the calculus is either one of the letters used as sentential variables (we will call such formulas elementary) or a compound of these letters, of the signs employed as sentential connectives, and of the parentheses. We agree to place each elementary formula in one of two mutually exclusive and exhaustive classes K_1 and K_2. Formulas that are not elementary are placed in these classes pursuant to the following conventions:

i) A formula having the form $S_1 \vee S_2$ is placed in class K_2 if *both* S_1 and S_2 are in K_2; otherwise, it is placed in K_1.

ii) A formula having the form $S_1 \supset S_2$ is placed in K_2, if S_1 is in K_1 and S_2 is in K_2; otherwise, it is placed in K_1.

iii) A formula having the form $S_1 \cdot S_2$ is

placed in K_1, if *both* S_1 and S_2 are in K_1; otherwise, it is placed in K_2.

iv) A formula having the form $\sim S$ is placed in K_2, if S is in K_1; otherwise, it is placed in K_1.

We then define the property of being tautologous: a formula is a tautology if, and only if, it falls in the class K_1 no matter in which of the two classes its elementary constituents are placed. It is clear that the property of being a tautology has now been described without using any model or interpretation for the system. We can discover whether or not a formula is a tautology simply by testing its structure by the above conventions.

Such an examination shows that each of the four axioms is a tautology. A convenient procedure is to construct a table that lists all the possible ways in which the elementary constituents of a given formula can be placed in the two classes. From this list we can determine, for each possibility, to which class the non-elementary component formulas of the given formula belong, and to which class the entire formula belongs. Take the first axiom. The table for it consists of three columns, each headed by one of the elementary or non-elementary component formulas of the axiom, as well as by the axiom itself. Under each heading is indicated the class to which the particular item belongs, for each of the possible assignments of the elementary constituents to the two classes. The table is as follows:

p	$(p \vee p)$	$(p \vee p) \supset p$
K_1	K_1	K_1
K_2	K_2	K_1

The first column mentions the possible ways of classifying the sole elementary constituent of the axiom. The second column assigns the indicated non-elementary component to a class, on the basis of convention (i). The last column assigns the axiom itself to a class, on the basis of convention (ii). The final column shows that the first axiom falls in class K_1, irrespective of the class in which its sole elementary constituent is placed. The axiom is therefore a tautology.

For the second axiom, the table is:

p	q	$(p \vee q)$	$p \supset (p \vee q)$
K_1	K_1	K_1	K_1
K_1	K_2	K_1	K_1
K_2	K_1	K_1	K_1
K_2	K_2	K_2	K_1

The first two columns list the four possible ways of classifying the two elementary constituents of the axiom. The second column assigns the non-elementary component to a class, on the basis of convention (i). The last column does this for the axiom, on the basis of convention (ii). The final column again shows that the second axiom falls in class K_1 for each of the four possible ways in which the elementary constituents can be classified. The axiom is therefore a tautology. In a similar way the remaining two axioms can be shown to be tautologies.

We shall also give the proof that the property of being a tautology is hereditary under the Rule of Detachment. (The proof that it is hereditary under the Rule of Substitution will be left to the reader.) Assume that any two formulas S_1 and $S_1 \supset S_2$ are both tautologies; we must show that in this case S_2 is a tautology. Suppose S_2 were not a tautology. Then, for at least one classification of its elementary constituents, S_2 will fall in K_2. But, by hypothesis, S_1 is a tautology, so that it will fall in K_1 for all classifications of its elementary constituents—and, in particular, for the classification which requires the placing of S_2 in K_2. Accordingly, for this latter classification, $S_1 \supset S_2$ must fall in K_2, because of the second convention. However, this contradicts the hypothesis that $S_1 \supset S_2$ is a tautology. In consequence, S_2 must be a tautology, on pain of this contradiction. The property of being a tautology is thus transmitted by the Rule of Detachment from the premises to the conclusion derivable from them by this Rule.

One final comment on the definition of a tautology given in the text. The two classes K_1 and K_2 used in the present account may be construed as the classes of true and of false statements, respectively. But the account, as we have just seen, in no way depends on such an interpretation, even if the exposition is more easily grasped when the classes are understood in this way.

Brief Bibliography

CARNAP, RUDOLF — Logical Syntax of Language, London, 1937.

FINDLAY, J. — "Goedelian sentences: a non-numerical approach," *Mind*, Vol. 51 (1942), pp. 259–265.

GÖDEL, KURT — "Über formal unentscheidbare Sätze der Principia Mathematica und verwandter Systeme I," *Monatshefte für Mathematik und Physik*, Vol. 38 (1931), pp. 173–198.

KLEENE, S. C. — Introduction to Metamathematics, New York, 1952.

LADRIÈRE, JEAN — Les Limitations Internes des Formalismes, Louvain and Paris, 1957.

MOSTOWSKI, A. — Sentences Undecidable in Formalized Arithmetic, Amsterdam, 1952.

QUINE, W. V. O. — Methods of Logic, London, 1950.

ROSSER, BARKLEY — "An informal exposition of proofs of Gödel's theorems and Church's theorem," *Journal of Symbolic Logic*, Vol. 4 (1939), pp. 53–60.

TURING, A. M. — "Computing machinery and intelligence," *Mind*, Vol. 59 (1950), pp. 433–460.

WEYL, HERMANN — Philosophy of Mathematics and Natural Science, Princeton, 1949.

WILDER, R. L. — Introduction to the Foundations of Mathematics, New York, 1952.

Index